On-Chip Pretreatment of Whole Blood by Using MEMS Technology

Edited by

Xing Chen

Chinese Academy of Sciences
China

eBooks End User License Agreement

CONTENTS

CHAPTERS

FOREWORD

With the cross-disciplinary development of MEMS technology, analytical chemistry, life sciences, electronics, informatics and so on, microfluidics technology has been in progess, which has been valued as one of the most important inventions affecting the human future. Blood is a very rare treasure, which has been widely used in clinic diagnoses, medicinal therapy, life sciences research *etc*. Blood is deemed so precious that is also called "red gold" because it can be sold for more than the cost of the same weight in gold. However the components of the blood is complicated, including cells, sub-cells, nucleic acid, protein, glucose, mineral ions, hormones, platelets and so on, which couldn't be directly analyzed or used but need be pre-treated. Traditional sample pre-treatment techniques are usually performed manually, at lower integration, with time-consuming processes, and requiring large volumes of blood. The technologies of blood sample pre-treatment microfluidic chip could not only resolve the above problems of the traditional techniques, but also could potentially enable whole blood analyses for personal diagnoses and point-of-care detection. Furthermore, the technologies of blood sample pre-treatment chip could play an important role for the revolution of modern medicine.

The group of Prof. Dafu Cui has begun to study MEMS technology and microfluidics. After more than ten years, various functional microfluidic chips such as blood separation chip, DNA chip, PCR chip, PDMS sandwich type electrophoresis chip have been developed and published in many famous international journals. As soon as "DNA extraction microfluidic chip integrated with porous matrix" was published, it was highly concerned and evaluated by the French and American experts. "Whole blood sample pre-treatment integrated microfluidic chip" has been cited so many times after it was published, which was highly valued as "This is a significant step toward a true "lab on a chip" where multiple steps are combined into a single device".

Blood sample pre-treatment microfluidic chip and its technology is on the rise, which is far from universal. The authors reviewed a large amount of literature, and combined ten years of the group's research work to write the eBook with rich, full of examples. The readers will get a comprehensive understanding of this new technology, and also be provided the reference value for their research work. Furthermore the blood sample pre-treatment microfluidic chip technology will be promoted and further developed in the future.

Prof. Daoben Zhu

Institute of Chemistry
Chinese Academy of Sciences

PREFACE

The Microfabrication technology has stimulated a plurality of lab-on-a-chip research for biomedical researchers and health care practitioners to manipulate and analyze complex biological fluids at the nano and microliter scale. Specifically, various miniaturized diagnostic devices have been developed for blood sample detection, which can increase diagnostic capacity significantly by enabling rapid, point-of-care chemical analysis. Recently, more and more researchers are paying attention to on-chip whole blood sample pretreatment, which is the focus of this eBook, aiming to form real Micro Total Analysis Systems (uTAS) for whole blood analysis by integrating blood sample pre-treatment and detection together.

The eBook consists of six chapters. Chapter **1** gives a brief introduction on the background of whole blood and the concept of μTAS, followed by a summary of specific microfluidic chips for whole blood pretreatment. Chapter **2** firstly reviews microfluidic chips for plasma isolation from whole blood samples with various methodologies (*i.e.*, microfiltration, microcentrifugation, ultrasound, Zweifach–Fung effect and capillary effect). Then two examples using the plasma skimming effect and crossflow filtration respectively are discussed in detail. Chapter **3** firstly reviews microfluidic devices for blood cell filtrating, sorting and collection. Sorting methods such as size filtration, optical and magnetic manipulation, affinity separation, dielectrophoretic manipulation are covered and compared. Then a case study using crossflow filtration is analyzed in detail. Chapter **4** firstly reviews reported microfluidic devices for blood cell lysis in which mechanical lysis, sonication lysis, thermal lysis, electrical lysis, chemical lysis and optical lysis are covered. And then a cell lysis microchip using chemical agents for cell lysis is chosen as an example for further concept demonstration. Chapter **5** firstly reviews microfluidic devices for DNA extraction and purification by comparing reported methods of solid phase extraction, solid phase reversible immobilization and liquid/liquid extraction. And then two microchips with different solid phase matrixes are used for demonstration of chip based DNA extraction. Chapter **6** firstly reviews current integrated microfluidic platforms for blood pretreatment, followed by a case study.

Both fundamental analysis and specific applications are presented in this eBook, which is written for researchers, engineers, and advanced students who have research interest in whole blood pretreatment and miniaturized techniques.

We would like to express our appreciation to Bentham Science Publishers and their team members.

Xing Chen
Chinese Academy of Sciences
China

List of Contributors

Xing Chen

State Key Laboratory of Transducer Technology, Institute of Electronics, Chinese Academy of Sciences, Beijing, 100190, China; E-mail: xchen@mail.ie.ac.cn; chenxing2004star@yahoo.com.cn

Dafu Cui

State Key Laboratory of Transducer Technology, Institute of Electronics, Chinese Academy of Sciences, Beijing, 100190, China; E-mail: dfcui@mail.ie.ac.cn

Jian Chen

State Key Laboratory of Transducer Technology, Institute of Electronics, Chinese Academy of Sciences, Beijing, 100190, China; E-mail: chenjian@mail.ie.ac.cn

CHAPTER 1

Introduction of Microfluidic Chips Targeting Whole Blood Pretreatment

Xing Chen*, Dafu Cui and Jian Chen

State Key Laboratory of Transducer Technology, Institute of Electronics, Chinese Academy of Sciences, Beijing, China

Abstract: The development of micro-total analysis systems targeting blood sample detections requests the availability of microfluidic devices for whole blood sample pretreatment including blood cell and plasma separation, white blood cell lysis and DNA purification, to name a few. In this chapter, background introduction of whole blood samples is conducted followed by a summary of key features and historical development of microfluidics. The marriage of microfluidics and whole blood pretreatment is highlighted with advantages and potential concerns discussed in detail.

Keywords: Microfluidics, sample pretreatment, whole blood.

1. WHOLE BLOOD COMPOSITION AND FUNCTION

Human blood is a highly complex substance. Its major components are Red Blood Cells (RBCs), which carry oxygen from the lungs to the body tissues; White Blood Cells (WBCs), which have major roles in disease prevention and immunity; and platelets, which are key elements in the blood clotting process. These blood elements are suspended in blood plasma, a yellowish liquid that comprises about 55% of human blood.

Through the circulatory system, blood adapts to the body's needs. When you are exercising, your heart pumps harder and faster to provide more blood and hence oxygen to your muscles. During an infection, the blood delivers more immune cells to the site of infection, where they accumulate toward off harmful invaders. Blood analysis has been used as indicators for various diseases. For example, RBC count is an indicator for anemia and the number of WBCs increases in infection and tumors. In addition, the number of platelets indicates whether bleeding or clotting is likely to occur.

Another function of blood lies in blood transfusion, which make modern medicine possible. Blood loss in surgery or in traumatic accidents can be replaced, allowing life-sustaining procedures such as open-heart surgery and organ transplantation to take place. In addition, severe anemic conditions such as sickle-cell anemia, caused by the under-production or defective production of RBCs, can be managed by RBC transfusion. All of these functions make blood a precious fluid. Each year million units of blood components are transfused to patients who need them. Blood is deemed so precious that is also called "red gold" because the cells and proteins it contains can be sold for more than the cost of the same weight in gold.

The characteristics and functions of the components in the whole blood are described in detail as follows: Plasma is the transporting medium for a myriad of hormones, electrolytes, sugars, waste products, and other substances. It is especially useful in transfusion medicine, as it provides the starting material for the preparation of critical blood-clotting factors, albumin and immune protein preparations. The clotting factor concentrates, prepared from large batches of pooled plasma, provide life-saving treatment for blood clotting disorders such as hemophilia.

RBCs are the most common type of cell found in the blood, with each cubic millimeter of blood containing 4-6 million cells, making up 40-45 percent of one's blood, and they give blood its characteristic color. With

*Address correspondence to Xing Chen: State Key Laboratory of Transducer Technology, Institute of Electronics, Chinese Academy of Sciences, Beijing 100190, China; Phone and Fax: +86-10-58887188; E-mail: xchen@mail.ie.ac.cn

a diameter of only 6 μm, RBCs are small enough to squeeze through the smallest blood vessels. In humans, as in all mammals, the mature RBC lacks a nucleus. This allows the cell more room to store hemoglobin, the oxygen-binding protein, enabling the RBC to transport more oxygen from the lungs throughout the body. RBCs are also biconcave in shape which increases their surface area for the diffusion of oxygen across their surfaces.

If a patient has a low level of hemoglobin, a condition called anemia, they may appear pale because hemoglobin gives RBCs, and hence blood, their red color. They may also tire easily and feel short of breath because of the essential role of hemoglobin in transporting oxygen from the lungs to wherever it is needed around the body.

WBCs (leukocytes) come in many different shapes and sizes, which are the body's mobile warriors in the battle against infection and invasion. Some cells have nuclei with multiple lobes, whereas others contain one large, round nucleus. Some contain packets of granules in their cytoplasm and so are known as granulocytes.

There are three types of WBCs: granulocytes, lymphocytes, and monocytes. There are, in turn, three kinds of granulocytes: neutrophils, eosinophils, and basophils. Neutrophils kill invading bacteria by ingesting and then digesting them. Eosinophils kill parasites, and are involved in allergic reactions. Basophils also function in allergic reactions, but are not well understood. Lymphocytes are key parts of the body's immune system. There are two kinds of lymphocytes: T cells and B cells. T cells direct the activity of the immune system. B cells produce antibodies, which destroy foreign bodies. Monocytes, the largest kind of WBCs, enter the tissues of the body and turn into even larger cells called macrophages to eat foreign bacteria and destroy damaged, old, and dead cells of the body itself.

Despite their differences in appearance, all of the various types of WBCs have a role in the immune response. They circulate in the blood until they receive a signal that a part of the body is damaged. Signals include interleukin 1, a molecule secreted by macrophages that contributes to the fever of infections, and histamine, which is released by circulating basophils and tissue mast cells, and contributes to allergic reactions. In response to these signals, the WBCs leave the blood vessels by squeezing through holes in the blood vessel walls. They migrate to the source of the signal and help begin the healing process.

Individuals who have low levels of WBCs may have more and worse infections. Depending upon which type of WBC is missing, the patient is at risk for different types of infection. For example, macrophages are especially good at swallowing bacteria, and a deficiency in macrophages leads to recurrent bacterial infections. In contrast, T cells are particularly skilled in fighting viral infections and a loss of their function results in an increased susceptibility to viral infections.

The platelets also called thrombocytes help blood to clot. When bleeding occurs, platelets clump together to help form a clot, or scab, over the wound. In their "resting" state, platelets look like two plates stuck together. When activated to form a clot, they change shape and look like tiny roundish blobs with tentacles. At only two to three microns, they are the smallest kind of blood cell.

Patients with malignant illnesses such as leukemia may have an insufficient number of platelets for effective clotting due to the disease itself or the harsh treatments necessary to combat these diseases. Transfusion of blood platelets now permits these patients to complete aggressive treatments without the risk of serious bleeding episodes.

2. MICROFLUIDICS

Microfluidics is the science and technology of systems that process or manipulate small (10^{-9} to 10^{-18} liter) amounts of fluids for a series of complex biochemical analysis [1]. They are constructed of various miniaturized biochemical analytical units, which are connected to each other *via* microchannels with dimensions of tens to hundreds of micrometers on solid substrates (glass, plastic, or silicon) by means of Micro Electro Mechanical Systems (MEMS) or other micromachining technologies.

The birth of microfluidic chips can be traced back to 1975 when Terry and his colleagues fabricated the first micro gas chromatography chip on a silicon wafer, enabling the separation of gas mixtures in a few seconds. In 1990, Manz [2] proposed the concept of miniaturized total chemical analysis systems (μTAS) and "officially" opened the hot research topic of microfluidics. The concept was initially put forward for new chemical sensors with enhanced performance. By and by, researchers realized that the reduction in device size brought several unique advantages verified by theoretical analysis and experimental results such as lower reagent consumption, faster analysis and lower cost.

Since 1992, various microfluidic chips were reported by a variety of researchers in both academia and industry, which mainly focused on the application of microfluidics in the area of electrophoresis for developing high performance analyzing systems such as gas chromatography, high performance liquid chromatography and capillary electrophoresis. Nowadays, a miniaturized total analysis system may consist of sample pretreatment modules [3, 4], biochemical reaction modules (PCR, DNA hybridization or immunoassay) [5-7], separation and detection modules (capillary electrophoresis) [8] and integrated multifunctional microchips [9-11], aiming for high-speed and high-throughout tests of inorganic ions, organic ions, proteins, nucleic acids, cells *etc.*

As a novel technology, microfluidics offers many advantages such as very small quantities of samples and reagents usage, separations and detections with high resolution and sensitivity, low cost, short times for analysis, and small footprints for the analytical devices. Small volumes, the key feature of microfluidics, reduce the time taken to synthesize and analyze a product, and also cut down reagent costs and the amount of chemical waste. It offers fundamentally new capabilities in the control of concentrations of molecules in space and time.

Following the current trend, microfluidics will be a practical technology widely used in a number of fields such as clinic medical diagnosis, drug discovery and environment monitoring in the near future. Many developed countries and world renounced companies (*e.g.*, Agilent, Hitachi, Caliper) have invested heavily in exploiting microfluidics for commercialization. From the perspective of academia, with the development of microfluidics, a distinct new field of BioMEMS has emerged, which becomes one of the most rapid developing fields in MEMS.

At the same time, microfluidics has suffered from several problems and disadvantages, which present challenges for its further development: (1) The effective integration of all essential elements of microchemical "factories" including the microchannels that serve as pipes, and other structures that form valves, mixers and pumps on a single chip is still under development; (2) The detection element is still commonly accomplished by conventional macrocouter parts; (3) The on-chip sample pretreatment component is still missing for the majority of microfluidic chip systems.

3. MICROFLUIDIC CHIPS FOR WHOLE BLOOD PRETREATMENT

As mentioned above, blood is a treasure of information about the functioning of the whole body. Every minute, the entire blood volume is recirculated throughout the body, delivering oxygen and nutrients to every cell and transporting products from and toward all different tissues. At the same time, cells of the immune system are transported quickly and efficiently through blood, to and from every place in the body where they perform specific immuno-surveillance functions.

Since, blood harbors a massive amount of information about the functioning of all tissues and organs in the body, blood sampling and analysis are of prime interest for both medical and science applications, and hold a central role in the diagnosis of many physiologic and pathologic conditions.

Since whole blood is very complicated on components, it is almost impossible to directly obtain this wealth of information for clinical and scientific applications without conducting sample pretreatment.

The golden standard procedure for conventional whole blood sample pretreatment is to spin a sample in a centrifuge to separate the blood sample into its individual components. The force of the spinning causes denser elements to sink, and further processing enables the isolation of a particular protein or the isolation of a particular type of blood cell. With the use of this traditional method, antibodies and clotting factors can be harvested from the plasma to treat immune deficiencies and bleeding disorders, respectively. Likewise, RBCs can be harvested for blood transfusion. However, this conventional method suffers from several drawbacks shown as follows:

For every leukocyte in blood which is most often the target of analysis there are a thousand more RBCs with it which need to be removed. Gentle extraction of the identified target cells from the complex blood sample is an even greater challenge. Because most of the leukocytes are designed to respond quickly to changes in their environment, they can easily be altered by the handling procedures during the separation steps. Several studies document that the exposure of cells to improper stimuli during the blood processing steps can alter the original immuno-phenotype of the separated cells.

Another challenge on whole blood sample pretreatment using the centrifugation force is the issue of biocompatibility. Whole blood samples are very complicated viscous mixtures including plasma, blood cells and blood platelets. The inner surfaces of the separation instrument contacting with blood samples should be inert and couldn't react with any component of the blood except in some special applications.

Microfluidic chips for whole blood pretreatment are attractive alternatives due to precise control over the cell microenvironment during separation procedures and the capability to scale down the analysis to very small volumes of blood. On the whole, on-chip blood sample pretreatment could lead to more gentle, fast, and consistent manipulation of the living cells, and therefore more accurate and better quality of extracted information. Furthermore, on-chip blood sample pretreatment is an essential component for the the development of micro-total analysis systems targeting blood sample detections. Fig. **1** summarizes step-by-step procedures of whole blood sample pretreatment using microfluidic chips.

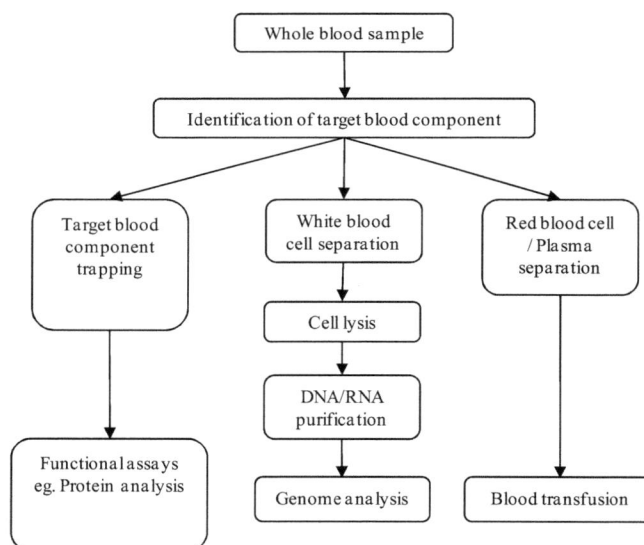

Figure 1: Schematic of the major steps in on-chip whole blood sample pretreatment.

A brief introduce of whole blood and microfluidic is covered in Chapter **1**. An overview of the ensuing chapters is as follows: On-chip isolation plasma from whole blood *via* microfluidic chips is covered in Chapter **2**. Blood cell separation and collection using microfluidic chips is introduced in Chapter **3**. On-chip blood cell lysis is summarized in Chapter **4**. Chapters **5** and **6** describe DNA purification leveraging MEMS technologies. All the figures in case studies of this eBook are provided by the author's group. For figures published previously, reprint with permission has been granted.

ACKNOWLEDGEMENT

The authors greatly acknowledge the financial support from the National Science Foundation of China under the Grant numbers of 60701019 and 60501020.

REFERENCES

[1]   Whitesides G. The origins and the future of microfluidics. Nature 2006; 442(7101): 368-73.

[2]   Manz A, Graber N, Widmer H. Miniaturized total chemical analysis systems: a novel concept for chemical sensing. Sensors and actuators B: Chemical 1990; 1(1-6): 244-8.

[3]   Lichtenberg J, de Rooij N, Verpoorte E. Sample pretreatment on microfabricated devices. Talanta 2002; 56(2): 233-66.

[4]   Mello A, Beard N. Dealing with 'real' samples: sample pre-treatment in microfluidic systems. Lab Chip 2003; 3(1): 11-20.

[5]   Liu J, Enzelberger M, Quake S. A nanoliter rotary device for polymerase chain reaction. Electrophoresis 2002; 23(10): 1531-6.

[6]   Yakovleva J, Davidsson R, Lobanova A, *et al.* Microfluidic enzyme immunoassay using silicon microchip with immobilized antibodies and chemiluminescence detection. Anal Chem 2002; 74(13): 2994-3004.

[7]   Lenigk R, Liu R, Athavale M, *et al.* Plastic biochannel hybridization devices: a new concept for microfluidic DNA arrays. Anal Biochem 2002; 311(1): 40-9.

[8]   Emrich C, Tian H, Medintz I, Mathies R. Microfabricated 384-lane capillary array electrophoresis bioanalyzer for ultrahigh-throughput genetic analysis. Anal Chem 2002; 74(19): 5076-83.

[9]   Woolley A, Hadley D, Landre P, Mathies R, Northrup M. Functional integration of PCR amplification and capillary electrophoresis in a microfabricated DNA analysis device. Anal Chem 1996; 68(23): 4081-6.

[10]   Waters L, Jacobson S, Kroutchinina N, Khandurina J, Foote R, Ramsey J. Microchip device for cell lysis, multiplex PCR amplification, and electrophoretic sizing. Anal Chem 1998; 70(1): 158-62.

[11]   Burns M, Johnson B, Brahmasandra S, *et al.* An integrated nanoliter DNA analysis device. Science 1998; 282(5388): 484.

Microfluidic Chips for Plasma Isolation from Whole Blood

Xing Chen[*], Dafu Cui and Jian Chen

State Key Laboratory of Transducer Technology, Institute of Electronics, Chinese Academy of Sciences, Beijing, P.R. China

Abstract: Since most conventional clinical assays are performed on cell-free serum or plasma, plasma isolation is a necessary step in the processing of whole blood sample pretreatment. Traditional technologies for plasma isolation are based on centrifugation or membrane filtration, which are time and cost intensive processes. Microfluidic chips for whole blood plasma isolation are attractive alternatives, which lead to more gentle, fast, and consistent manipulation of the whole blood samples, and therefore more accurate and better quality of plasma isolation. In this chapter, a variety of microfluidic devices for blood plasma separation are reviewed and compared. Two specific examples of plasma isolation leveraging MEMS technologies are included for further concept demonstration in which design, fabrication and testing of these microfluidic devices are covered in detail. These plasma isolation microfluidic devices can be easily connected with other microfluidic components to form micro total analysis systems for blood testing with the advantages of continuous and real time separation of plasma from whole blood samples.

Keywords: Plasma isolation, nano/micro structures, MEMS, microfluidics.

1. INTRODUCTION

Plasma is the transporting medium for a myriad of hormones, electrolytes, sugars, waste products, and other substances. It is especially useful in transfusion medicine, as it provides the starting material for the preparation of critical blood-clotting factors, albumin and immune protein preparations. The clotting factor concentrates, prepared from large batches of pooled plasma, provide life-saving treatment for blood clotting disorders such as hemophilia.

Separation of plasma from the whole blood for blood tests is a necessary step, since clinical diagnostics are often performed on cell free serum or plasma instead of raw blood to avoid the potentially side effects of blood cells and cellular components, which may lead to difficulties in analysis standardization.

Isolated blood plasma can also be used for plasma exchange treatment, plasma donation and cytapherasis such as platelet apheresis, medical diagnostics and blood biochemical analyses *et al.* [1, 2].

Traditional technologies for plasma isolation are based on centrifugation or membrane filtration, which are time and cost intensive processes. Microfluidic chips for whole blood plasma isolation are attractive alternatives, which lead to more gentle, fast, and consistent manipulation of the whole blood samples, and therefore more accurate and better quality of plasma isolation.

In this chapter, the traditional methods of blood plasma separation are introduced as background introduction. And then the current miniaturized methods for blood plasma isolation are reviewed, followed by two case studies to demonstrate how to design, fabricate and apply these microfluidic devices for plasma separation from whole blood samples.

2. TRADITIONAL METHODS

Plasma isolation is defined as the separation of blood cells from plasma, which belongs to particle-liquid

*Address correspondence to Xing Chen: State Key Laboratory of Transducer Technology, Institute of Electronics, Chinese Academy of Sciences, Beijing 100190, P.R. China; Tel and Fax: +86-10-58887188; E-mail: xchen@mail.ie.ac.cn

separation and can be achieved by subjecting the suspension to natural or artificially induced gravitational fields. Blood cells are denser than the liquid medium of plasma in which they are suspended. These cells would settle down due to gravitational forces and form a zone with very high blood cell density. This zone is referred to as the sediment and the clear liquid left behind is referred to as the supernatant. This principle is utilized in separation processes such as sedimentation and centrifugation.

It is worth noting that the method of centrifugation is used not only for plasma isolation but also for different blood cell separation [3-8].

Another useful and important method for plasma isolation is the membrane filtration technique by forcing the suspension through a porous membrane which allow the liquid to go through while retain the particles [9-14]. Various membranes were also designed and fabricated for blood cell separation [15-20].

The membrane based filtration technique is widely applied to the collection of plasma by plasmapheresis, a procedure in which plasma is separated from whole blood, and the cellular components are reinfused to the donor. Although still somewhat controversial, there is a general understanding of the physics and fluid mechanics involved in the transport phenomena associated with the microporous filtration of whole blood. The filtration process is governed by concentration polarization, with the red blood cell being the polarizing species. Shear-enhanced cellular diffusion has been proposed to explain how such a large cell can depolarize to allow the observed separation rates. The filtration rate per unit area is proportional to the first power of the shear rate and therefore, can remain approximately constant if the shear rate is increased to accommodate a decrease in surface area.

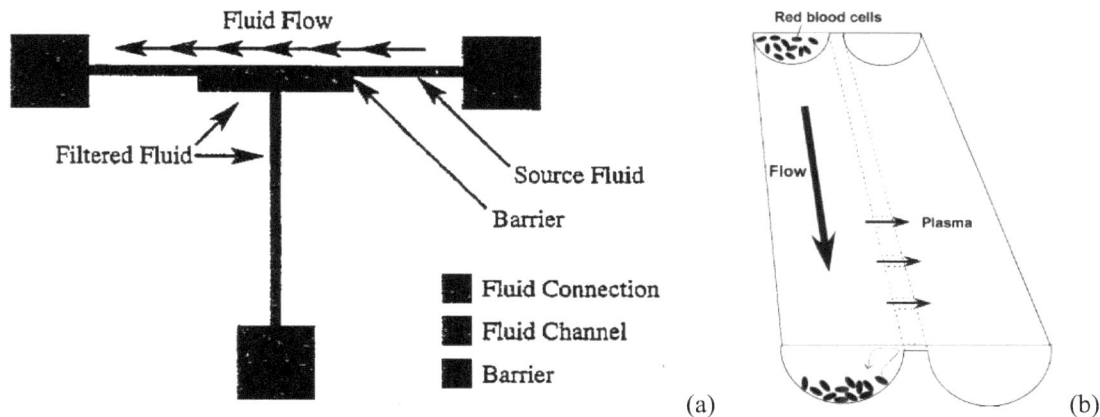

Figure 1: (a) A schematic drawing of the filter design [21]. The fluid enters the device from the top right port and flows to the exit port at left. Some particle-free fluid is pushes across the barrier and down to the filtered fluid exit port. By controlling the pressure drop between the top right and bottom ports, we can control how quickly and how much filtered fluid is drawn out. Shear forces inside the horizontal channel act to prevent clogging of the barrier by particles. Reprinted with permission from Elsevier. Copyright (1996) (b) Principle of plasma separation from whole blood in microchannels [22]. Reprinted with permission from the American Chemical Society. Copyright (2009).

Three automated membrane-based plasmapheresis systems are currently in use, primarily at fixed-site locations [23-26]. Two use hollow fiber membrane filter modules, while the third uses a rotating cylinder which relies on Couette rather than Poiseuille flow to generate the required shear rates at the membrane surface.

A number of additional membranes and filtration modules of different designs have been investigated as potential candidates for donor plasmapheresis [27-29]. Some are being used successfully in therapeutic plasma exchange systems [30, 31].

Two portable systems have been developed; one centrifuge-based system requires house current, the other membrane system uses gravity, augmented by a battery power source [32-35]. These two systems, due to their small sizes and low weights, are well suited for use in environments where the equipment has to be moved on a frequent basis.

However, traditional techniques are usually performed manually, at macroscale, with multi-step processes, and require large volumes of blood and skilled operators.

3. CURRENT MICROFLUIDIC DEVICES TARGETING BLOOD PLASMA ISOLATION

With the development of lab-on-chip technologies, chemical and biological experiments can be performed on all-in-one, automatic lab-on-chip devices, which require fewer reagents, lower sample volume and a shorter assay turnaround time. On-chip blood separation of diluted blood or raw blood has been implemented by several groups using various principles, shown as follows.

3.1. Microfiltration

Filtration is based on particle size differences (Red Blood Cells (RBCs) are 2µm in thickness whereas White Blood Cells (WBCs) are larger than 5µm) and selective segregation by dimensioned pores or filtration microstructures. This method makes it easy to extract the liquid phase of a suspension so that it is commonly used for blood plasma separations with some successful designs [36].

Wilding [37, 38] and He [39] fabricated an array of posts inside a microchannel as microfilters for cell/ particle separation (details showed in next chapter). While these approaches are capable of providing reusable filters, they are incapable of separating plasma from whole blood samples since the filtration gaps between the posts are larger than red blood cells.

Another method used for microfilters is to fabricate packed bed columns, either by packing beads [40] or by *in situ* polymerization [41].

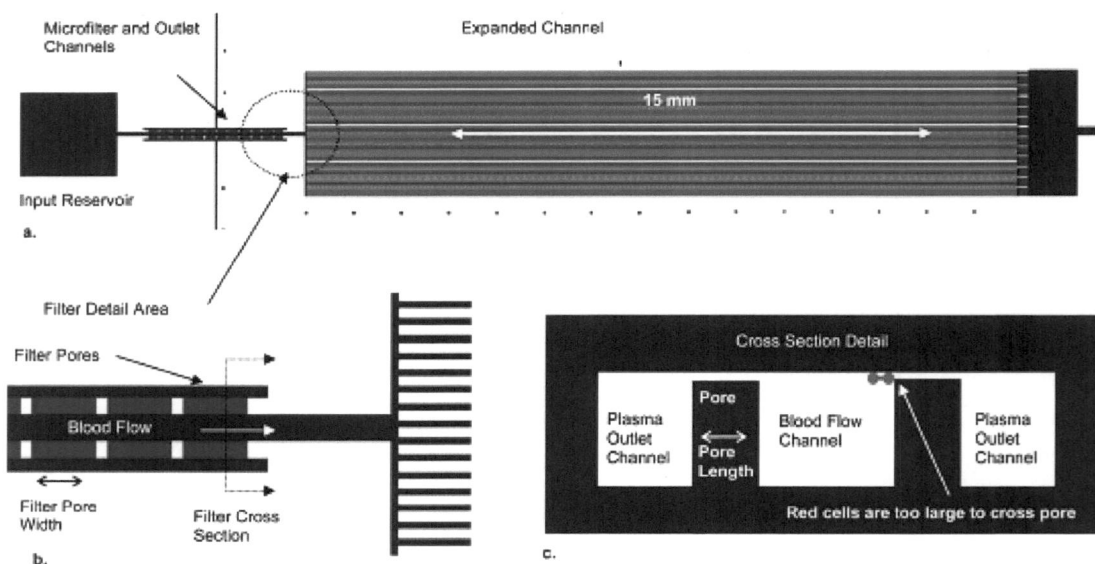

Figure 2: Microfilter device design and detail [42]: (**a**) top view of generic device design with narrow and expanded channels, (**b**) filter detail area showing filter pores and expanded channel layout, (**c**) microfilter cross section. Reprinted with permission from the Royal Society of Chemistry. Copyright (2005).

Moorthy reported porous microfilters based on emulsion polymerization for plasma separation fabricated inside a microchannel for blood separation [43]. The fabrication of the device was a multi-step process with *in situ* polymerization to form a porous plug that is permeable for fluid but impenetrable for particles or cells. The porosity of the filter is dependent on the composition of the pre-polymer mixture. The efficiency of plasma separation by the porous filter was tested with comparable performance of centrifugation. The porous microfilter thus mimics the functionality of a centrifuge, with additional benefits of no power consumption and ability to handle small sample volumes.

Thorslund [44] developed a microfliter for plasma isolation by using a commercially available membrane. The microfilter is a hybrid system containing a hydrophilic polypropylene membrane, a bottom chip substrate (I) with patterned microchannels and a lid substrate (II). The membrane filter was incorporated into the lid before the two layers of PDMS were bonded together. The PDMS substrates were casted from patterned silicon wafers, which were fabricated using standard MEMS techniques. The isolated plasma was collected in underlying microchannels based on hydrophilic effects and capillary forces in the channel system. Different membranes, such as polypropylene (PP), polycarbonate (PC) and polyvinylpyrrolidone/polyethersulfone (PVP/PES), were used to remove blood cells from the plasma. The homogenously pore sized polypropylene (PP) turned out to be the best candidate, which was compatible with a larger volume with less diluted whole blood without hemolysis or leakage.

However, in these reported dead-end microfilters, the entrances to the pores were quickly clogged [41, 43] due to the deformability of the blood cells, which make these microfilters with limited lifepans Dead-end filtrations may be efficient, but they request high sample dilution ratios, low flow rates and optimized geometries to alleviate the blockage issue. For dead-end filtration, the fluidic flow is perpendicular to the filtration structures so that smaller particles pass through the filtration barriers along the fluid, while larger particles are stopped to cause clogging or jamming. To overcome clogging of dead-end filters, crossflow filtration, which allows the bigger particles to stay in a suspended state instead of being aggregated around filter posts, was used in microfluidic chips in later development.

Brody *et al.* [21] firstly suggested the crossflow filtration using a planar microfabricated filter device, shown in Fig. **1 (a)**. The device was fabricated by using a three-mask-level process. The first level defines connection ports, which were etched completely through the wafer to the rear side of the silicon. The second level defines the fluid-transport channels, and the third level defines the maximum size of particles that could flow through the filter. 16 μm and 2.6 μm fluorescent spheres were used to quantify the performance of the device. Although this filter couldn't be used to separate plasma from whole blood samples, it has several unique features: In this microfliter, a sloped profile was used to provide a gradual transition from the narrow filter region to the wide flow channel, which reduces the pressure needed to overcome the surface tension. In addition, a tangential flow to reduce clogging was used and particles built up along the filter region were carried downstream due to tangential forces. Particles trapped around the filter area can be dislodged from the filter by backpressure easily.

Then Crowley *et al.* [42] developed a passive transverse-flow microfilter device, operating entirely on capillary action, for isolating nanoliter volumes of plasma from a single drop of blood. This transverse-flow microfilter (blood flow is parallel to the filter face) was micromachined in silicon and glass substrates, and used to study the engineering variables controlling the 'on-chip' separation of plasma from whole blood. As shown in Fig. **2**, the generic microfilter design consists of an input reservoir, narrow filtration channels with transverse flow microfilters, a plasma outlet channel to collect filtered plasma, and a wider expanded channel connected in series with the filtration channel. Flow channels are 10 μm deep, and the filtration channel is 100 μm wide. The expanded channel areas were fabricated as arrays of parallel flow channels which are 45 μm wide and 15 mm long.

The effects of blood shear rates and RBC volume fraction (hematocrit) on plasma filter flux were investigated and compared to well-known macroscale blood filtration results using microporous membranes. Similar to macroscale plasmapheresis operations, filter flux in this microdevice conforms to a power law model, with filter flux as a function of the wall shear rate of blood in the filtration channel. Unlike macroscale operations, microfilter plasma flux is insensitive to hematocrit levels between ~20 and 40%. But for a capillarity driven system with laminar flow, an accumulation of blood cells at the leading edge always results in substantial flow instabilities.

VanDelinder [46] furthered the idea of crossflow filtration with a glass-PDMS hybrid microchip, which is made of a single mold of a PDMS sealed with a cover glass. When loaded with blood diluted to 20% hematocrit and driven with pulsatile pressure to prevent clogging of the channels with blood cells, the device operated for at least 1 h, extracting 8% of blood volume as plasma at an average rate of 0.65 µL/min. The extracted plasma was capable of meeting the standards for common assays and was delivered to the device outlet 30 sec after blood injection at the inlet. Integration of the cross-flow microchannel array with on-chip assay elements has the potential to create a microanalysis system for point-of-care diagnostics, capable of reducing costs, turn-around times, and volumes of blood sample and reagents required for the assays.

Tachi [22] reported a microchip with an inter channel microstructure, shown in Fig. **1(b)**. The plasma separation was based on both cross-flow filtration and sedimentation of red blood cells in the microchannels. This microchip with an inter channel microstructure was used to simultaneously separate plasma from human whole blood and dilute the plasma without hemolysis in about 3 min. It is possible to collect plasma from just one or several drops of whole blood by using this microchip. This microchip and microstructure can be easily connected and integrated with the other microfluidic devices and microchannels for quantitative analyses.

3.2. Microcentrifugation

Another approach focusing on separating plasma from whole blood is based on centrifugal force. Centrifugation is an enhanced sedimentation technique, where gravity acceleration is supplemented by the acceleration of a rotating system. Centrifugal microfluidic platforms [47-51] are of particular interest for assay integration as their artificial gravity fields intrinsically implement pumping forces with established methods for particle separation.

The centrifugation effect has been used in two kinds of microfluidic devices for plasma extraction. On the one hand, Haeberle *et al.* [52] proposed the use of a decanting chamber on a rotating lab-disk, extracting 2 µL plasma from5 µL blood in 20 sec at a spinning rate of 40 Hz per second. This process is simple and can be integrated with on-disk analysis units, but it requires a standard rotary driver.

On the other hand, some researchers have utilized the centrifugal force induced by flow in curved channels for continuous fragmentation [45]. Two specific examples with different geometrical designs (a spiral and a 180° bend) are shown in Fig. **3 (a-c)**.

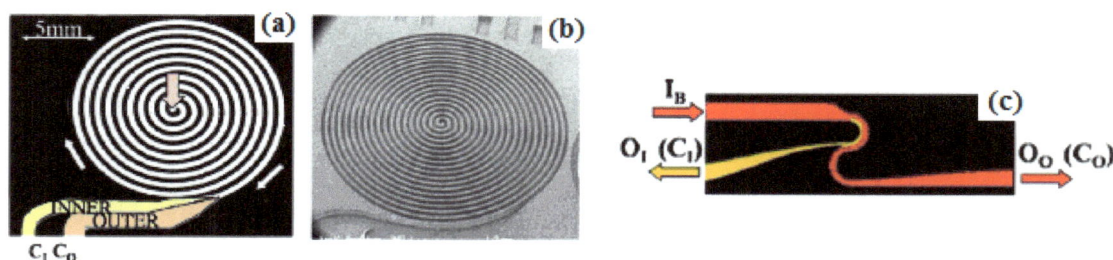

Figure 3: Centrifugation in spiral [45]. (**a**) Schematic of spiral. Injection is through the centre of the spiral. Channels are 200µm deep and 200µm wide. (**b**) Scanning electron microscope photographs of Spiral design. (**c**) Injection in a 180° bend. The curve is 200µm wide with a radius of curvature from 200 to 400µm. Reprinted with permission from Elsevier. Copyright (2009).

In the spiral design, the liquid is injected at the centre of the spiral through a deep etched hole. At the end of the spiral, the liquid is equally separated between its inner (I) and outer (O) outlets due to generated centrifugal accelerations (Fig. **3a** and **b**). The channels are 200 µm deep, 200 µm wide and 12 cm long. Several tests were performed with low concentrated 5 µm fluorescent beads suspended in plasma. By counting the fluorescent beads at each outlet (C1and C0), bead separation efficiency were estimated and linked to hydrodynamic centrifugal effects.

Zhang [53] reported a simple microchannel network on lab-on-CD for separating plasma and blood cells from the whole blood into two different reservoirs. In this design, only small droplets of blood were needed (typically ~1 μL). Compared to traditional sedimentation methods in which blood cells and plasma are separated into different layers within one tube/reservoir, in this lab-on-CD method blood cells and plasma flow are separated into two microchannel branches continuously. Therefore subsequent blood tests can be conducted in the microchannel branch of plasma immediately after separation. This merit could be very useful for fast, qualitative blood analysis for the purpose of prescreening pathogens.

This CD platform includes a microchannel network consisting of a straight main microchannel, a curved microchannel and a branched microchannel, shown in Fig. **4**. The CD-type microdevice was fabricated by the soft photolithography technique. Firstly, a microstructured master on SU-8 was obtained by the traditional photolithography technique. PDMS was then molded on this master with grooves molded into the PDMS structural material. A 1 mm diameter through hole was drilled using a blunt needle at the location of the inlet reservoir of the PDMS layer in order to perfuse the blood samples. A 0.5 mm diameter through hole was aligned to each outlet reservoir in order to balance the pressure. The microchannel network was formed after the patterned structural material (PDMS here) was bonded to a glass covers lip slide.

Up to 99% RBCs were separated and about 22% plasma was recovered from the diluted blood of 6% hematocrit by using this CD-type microdevice. The merits of this design are its simple structure, small operating time and high separation efficiency with blood separation within 1-2 sec. There is no specific requirement for the amount of blood. Therefore this method provides flexibility for drawing the blood samples, and for qualitative blood analysis, in particular, prescreening of pathogens.

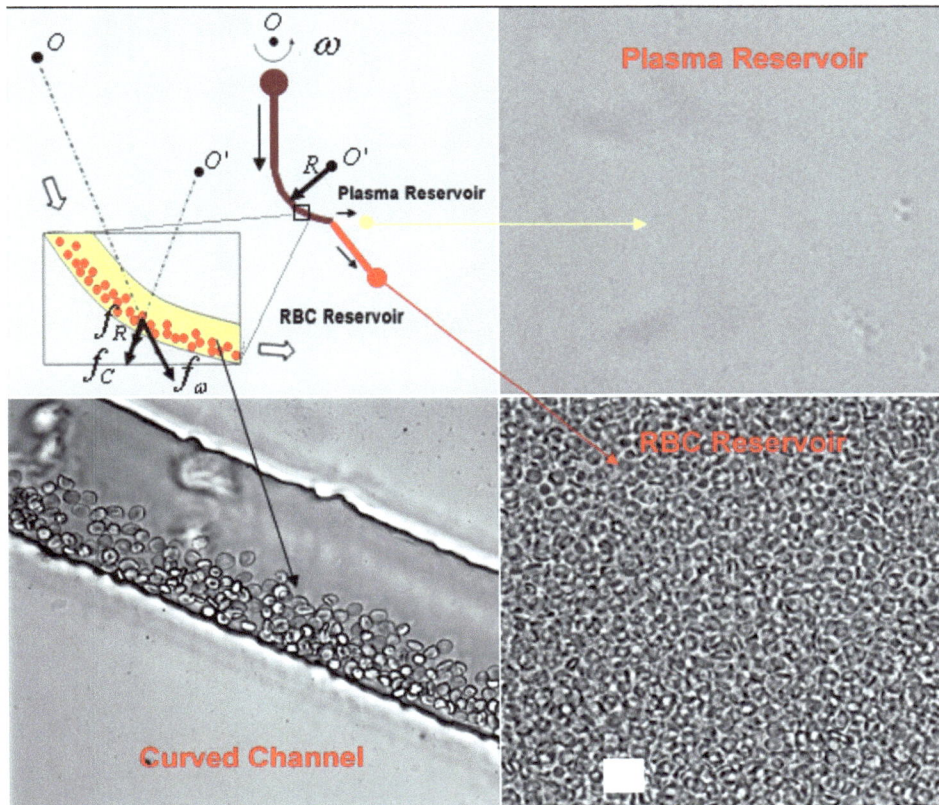

Figure 4: Separation mechanisms and separation results [53]. (**a**) Schematic diagram of the microchannel network which is not drawn to scale, (**b**) flowing blood is separated in the curved microchannel, (**c**) nearly pure plasma flows into the plasma reservoir and (**d**) RBCs are accumulated in the RBC reservoir. Reprinted with permission from the Institute of Physics. Copyright (2008).

Steigert [54] furthered this method of microcentrifugation with a fully integrated centrifugal microfluidic "lab-on-a-disk" for rapid colorimetric assays in human whole blood. All essential steps comprising blood sampling, metering, plasma extraction and the final optical detection were conducted within 150 sec in passive, globally hydrophilized structures. This structure obviates the need for intricate local hydrophobic surface patterning. Plasma was first centrifugally separated, and then metered by an overflow and subsequently extracted by a siphon-based principle through a hydrophilic extraction channel into the detection chamber. A hot embossed polymer substrate of the size of a conventional Compact Disc (CD) featuring fluidic and optical elements was used to perform colorimetric assays. The fluidic elements provide ports for sample and reagent uptake as well as a sample preparation structure for an integrated blood sedimentation and plasma metering which is connected to a combined mixing and detection chamber.

Lee [55] reported a portable, disc-based, and fully automated enzyme-linked immuno-sorbent assay (ELISA) system to test infectious diseases from whole blood, shown in Fig. **5**. The centrifugal microfluidic device was utilized for plasma separation and the full integration of microbead-based suspension ELISA assays on a disc starting from whole blood. The concentrations of the antigen and the antibody of Hepatitis B Virus (HBV), HBsAg and Anti-HBs respectively, were measured using the lab-on-a-disc (LOD). All the necessary reagents were preloaded on the disc and the total process of the plasma separation, incubation with target specific antigen or antibody coated microbeads, multiple steps of washing, enzyme reaction with substrates, and the absorbance detection were finished within 30 min. Compared to the conventional ELISA, the operation time was dramatically reduced from over 2 hours to less than 30 min while the limit of detection was kept similar.

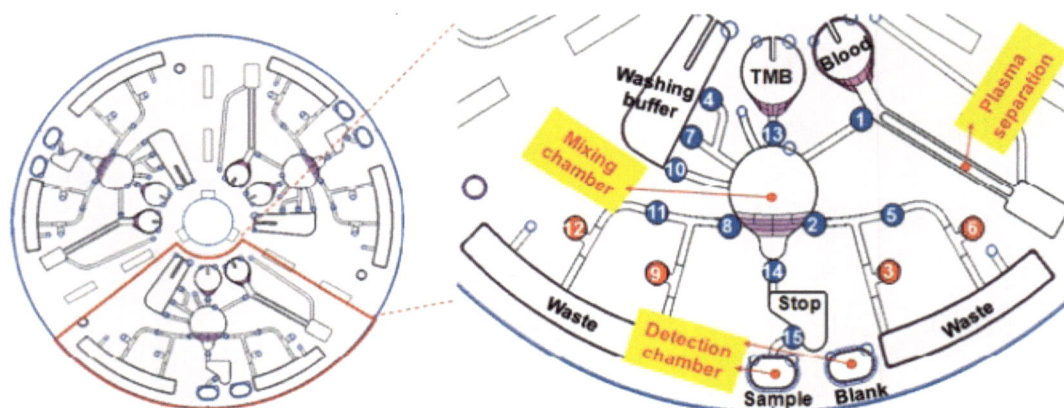

Figure 5: Disc design showing the detailed microfluidic layout and functions [55]. Reprinted with permission from the Royal Society of Chemistry. Copyright (2009).

3.3. Ultrasonic Force

Pressure fluctuations in a liquid medium result in acoustic radiation forces on suspended particles. As long as the diameter of the particles is much smaller than half the wavelength of the standing wave these forces can act mainly in one direction and the particles can move towards either a pressure node, a pressure anti-node or not at all [56].

Cousins [57] developed a technique by using the ultrasonic standing wave to efficiently separate plasma from human whole blood. 3-mL samples were held on the axis of a tubular transducer and exposed for 5.7 min to an ultrasonic standing wave. The cells were concentrated into clumps at radial separations of half wavelength. With the growing of the clumps in size and sediment under gravity, a distinct plasma/cell interface formed as the cells started to sedimente. At the same time, the volume of clarified plasma increased with time. There was no measurable release of haemoglobin or potassium into the suspending phase, indicating that there was no mechanical damage to cells at the specific frequency. A total of 114 samples from volunteers and patients were subsequently clarified in a 1.5-MHz system driven by an integrated generator with the average efficiency of clarification of blood as high as 99.76%. Although the tubular transducer of the ultrasonic standing wave was big, the technique of ultrasonic force was proved to be efficient for plasma isolation.

After Cousins reported a tube-type sorter, the group of Laurell [58, 59] furthered the method of ultrasonic standing waves for bioseparation. In this design, a miniaturized separator with a microchannel net was developed. Silicon with good acoustic properties was chosen to make the separation chips by using photolithography and anisotropic wet etching techniques. The wave has a node in the middle of the microchannel, leading to enrichment of RBCs in the middle and depletion of their concentration in the periphery of the microchannel. Plasma is partially separated from blood in the microchannel flow due to different acoustic impedances between blood cells and plasma.

The direction and size of the acoustic forces can be estimated theoretically and used to separate different components. If a micro channel is actuated at its fundamental resonance frequency, pressure anti-nodes are present along the side walls of the channel and a pressure node along the centre. When, suspended particles enter microchannels, they are affected by the acoustic forces and move towards either the pressure node (Fig. **6a**) or the pressure anti-nodes (Fig. **7a**) depending on the density and compressibility of the particles and the medium respectively. Since laminar flow conditions prevail (small channel dimensions and reasonably low flow rates), the particles remain in their lateral positions even though being outside the acoustic field region until they reach the end of the channel, which is split into three different outlets. The particles can subsequently be collected in the centre outlet (Fig. **6b**) or the two side outlets (Fig. **7b**). By controlling the flow rate through the three outlets independently the fraction of the medium that exits together with the particles can be controlled.

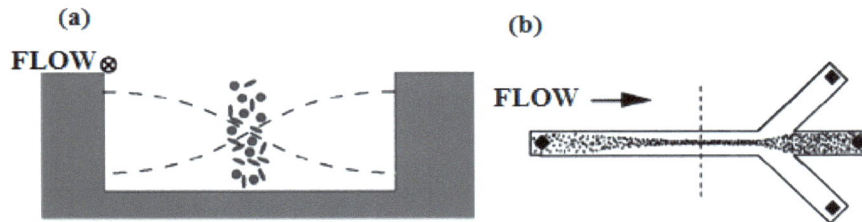

Figure 6: (a) Particles positioned, by the acoustic forces, in the pressure nodal plane of a standing wave. (Cross section of the channel in (**b**), dashed line.) (**b**) Top view of a continuous separation of particles, positioned in the pressure node, from a fraction of their medium [58, 59]. Reprinted with permission from the Royal Society of Chemistry. Copyright (2004).

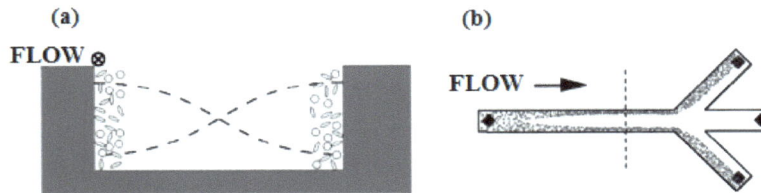

Figure 7: (a) Particles positioned, by the acoustic forces, in the pressure anti-nodal plane of a standing wave. (Cross section of the channel in (**b**), dashed line.) (**b**) Top view of a continuous separation of particles, positioned in the pressure anti-nodes, from a fraction of their medium [58, 59]. Reprinted with permission from the Royal Society of Chemistry. Copyright (2004).

Then the same of the researchers [60, 61] utilized acoustic forces to manipulate the suspended RBCs, enabling continuously translation of these blood cells in a laminar flow microchannel from one medium to another with virtually no mixing of the two media, shown in Fig. **8(a)**. RBCs were switched from blood, spiked with Evans blue, to clean blood plasma. At least 95% of the RBCs (bovine blood) were collected in clean blood plasma while up to 98% of the contaminant was removed. The obtained results indicate that the presented method can be used as a generic method for particle washing and, more specifically, be applied for plasma exchange therapy.

Then the same group of Laurell [62, 63] developed an acoustic whole blood plasmapheresis chip for prostate specific antigen microarray diagnostics. A new acoustophoresis chip was designed to separate plasma with high concentration blood cells, aiming to resolve the limitations of the previously presented microfabricated silicon acoustic separator chips, [59, 64], which were not able to process sufficiently high particle concentrations, partly due to the very high acoustic forces required to concentrate particles into a band sufficiently narrow to enable separation.

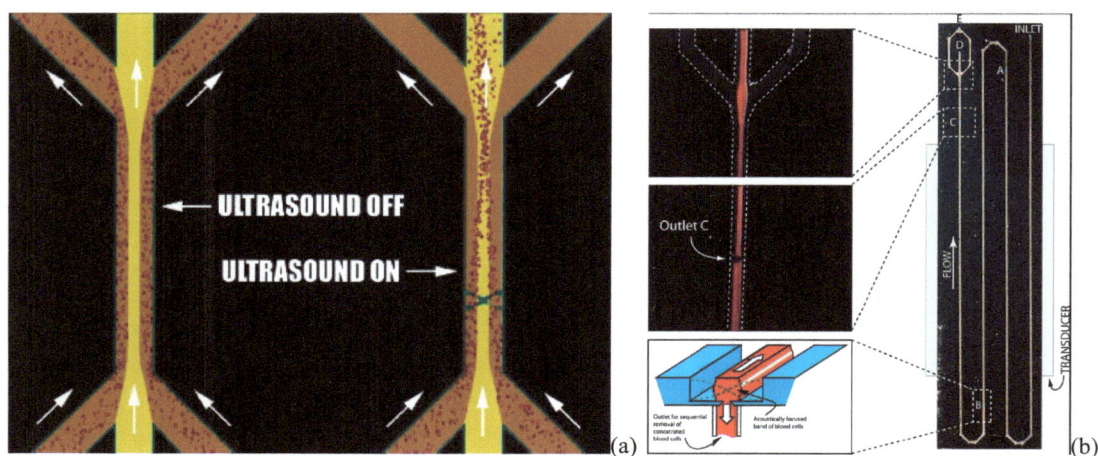

Figure 8: (**a**) Schematic illustration of the medium exchange principle. The particles in the contaminated medium enter through the side inlets and exit through the side outlets when the ultrasound is turned off (left). If the ultrasound is turned on, as indicated by the schematic standing wave in the right microfluidic structure, the particles are switched over to the clean medium and exit through the center outlet together with the clean medium, whereas the contaminated medium still flows to the side outlets [60, 61]. Reprinted with permission from the American Chemical Society. Copyright (2005); (**b**) Principle of plasmapheresis. Acoustic standing waves gather blood cells in the pressure node located in the middle of the separation channel. Enriched blood cell fractions are removed through outlets A-C, thus decreasing the hematocrit gradually in the channel. The remaining focused blood cells exit through outlet D while the clean plasma fraction is withdrawn from exit E. The transducer placed underneath the chip generates an ultrasonic standing wave between the channel walls, perpendicular to the flow. Dotted lines have been added to the microscope images (upper and middle left) to outline the channel boundaries [65]. Reprinted with permission from the American Chemical Society. Copyright (2009).

In the previously reported designs, the relatively short (15-30 mm) separation channel ends in a trifurcation where concentrated blood cells exit through the central outlet and clean plasma exit through the side branches. The new separator was modified as follows: first, the separation channel was elongated in a meander type of fashion, which enables the blood cells to be affected by the acoustic standing wave for a longer time period. Second, several extra outlets were added along the separation channel placed in the middle of the separation channel allowing the removal of blood cells already focused without removing a large part of the blood plasma. This procedure reduces the concentration of blood cells in the separation channel so that at the end of the separation channel all remaining cells can be packed in a sufficiently narrow band prior to reaching the flow splitter, enabling a higher separation efficiency.

The separation microfluidic chip was used to prepare diagnostic plasma from whole blood, demonstrating a strong link to clinical applications, shown in Fig. **8(b)** [65]. This acoustic separator has the capacity to sequentially remove enriched blood cells in multiple steps to yield high quality plasma of low cellular content.

3.4. Zweifach-Fung Effect, Lateral Migration

Yang [66-69] reported a microfluidic device for continuous, real time blood plasma separation, shown in Fig. **9**. The principle of the blood plasma separation from blood cells is supported by the Zweifach-Fung effect and was experimentally demonstrated using simple microchannels.

This blood plasma separation device was composed of a blood inlet, a bifurcating region which leads to a purified plasma outlet, and a concentrated blood cell outlet using an analogous electrical circuit, as well as analytical and numerical studies. It was designed to separate blood plasma from an initial blood sample of up to 45% inlet hematocrit (volume percentage of cells). During 30 min of continuous blood infusion through the device, all the RBCs traveled through the device toward the concentrated blood outlet while only the plasma was separated at the bifurcating regions and flowed towards the plasma outlet.

This device was capable of operating continuously without any clogging or hemolysis of cells. The experimentally determined plasma selectivity with respect to blood hematocrit level was almost 100% regardless of the inlet hematocrit. The total plasma separation volume percent varied from 15% to 25% with increasing inlet hematocrit. Due to the device's simple structure and control mechanism, this microdevice is expected to be used for highly efficient continuous, real time cell-free blood plasma separation from blood samples for use in lab.

Also based on the Zweifach-Fung bifurcation law, RBCs were separated from plasma flowing in microchannels reported by Kersaudy-Kerhoas [70, 71]. In this study, daughter channels were placed alongside a main channel such that cells and plasma were collected separately. The device was a versatile but yet very simple module producing high-speed and high-efficiency plasma separation.

In another paper [72], a unique microfluidic channel geometry called "corner-edge" was designed based on plasma skimming effect at branching channel. The clear plasma layer was enhanced and the efficiency of separation was above 99% comparing with whole blood hematocrit.

The working principle of the plasma skimming effect is quite similar as the Zweifach-Fung bifurcation law, which will be illuminated in the case study in detail.

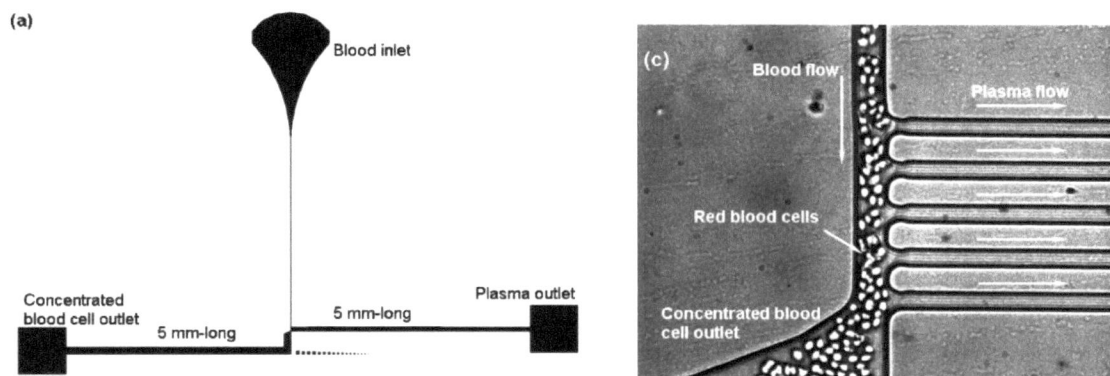

Figure 9: Schematic diagrams of a microfluidic blood plasma separation device [66-69]; (**a**) an overview of a device. This device is designed to have a whole blood inlet, a purified plasma outlet, and a concentrated blood cell outlet. Each channel is 5 mm in length. (**b**) A snapshot of same region taken using a higher resolution CCD camera with shorter shutter open time (20 ms). Reprinted with permission from the Royal Society of Chemistry. Copyright (2006).

3.5. Blood Rheology, Inertial Effect

The concept of hydrodynamic lift of deformable cells in shear flow and the cell-free layer produced adjacent to the boundaries is shown as follows. As one of the oldest and best known hemodynamic phenomena of the microcirculation the decreased hematocrit concentration in small tubes relative to the larger reservoirs to which they are connected [73] is arising from hydrodynamic drift of the deformable cells away from the high shear rate (and high shear stress) regions adjacent to the walls of the vessel [74-76]. Thus, a cell-free layer, typically comparable to the size of a single cell, exists in large portions of the microcirculation. This cell free layer is also responsible for the nonuniform distribution of cells in a vessel network such as a tissue. Briefly, the concentration of RBCs in a network of daughter arterioles, capillaries or venules in the microcirculation can be lower than the concentration in the mother feeding vessel; this is usually referred to as the plasma skimming [77, 78].

An innovative device for separation of cells from plasma was proposed by using the plasma skimming concept as a developmental strategy based on the cell-free layer and its local enhancement by geometric singularities [79]. A very narrow channel with a high shear stress was designed to enhance the downstream cell-free layer. The principle is purely passive and with no need for filters.

In this design, a rapid variation of the cross-section of a microfluidic channel was used to modify the spatial distribution of cells downstream of a constriction and increase the cell-free layer adjacent to the boundary. In (Fig. **10**), various designs of microfluidic devices for plasma isolation were shown, with three channels in parallel at the downstream of the constriction. Because of the enhanced cell-free layer, the liquid collected in the outermost channels is totally composed of plasma (Fig. **10a**). When the width of the outermost channels was broadened to extract more plasma some cells were also entrained (Fig. **10b**). In such cases, decreasing the width or increasing the length of the constriction again resulted in the extraction of essentially pure plasma as shown successively in (Fig. **10c** and **d**).

Based on this method, a maximum extraction of 10.7% was scored for 1/20 diluted blood, with excellent purity (red cell contamination is similar to that of centrifuged plasma) and a high injection flow rate. For all these reasons, this device seems promising even if the extraction yield is still too low. The next step is to enhance the cell-free layer further before it supplies the singularities, in order to shift the dilution limit and work on more concentrated samples, but above all to increase the extraction yield.

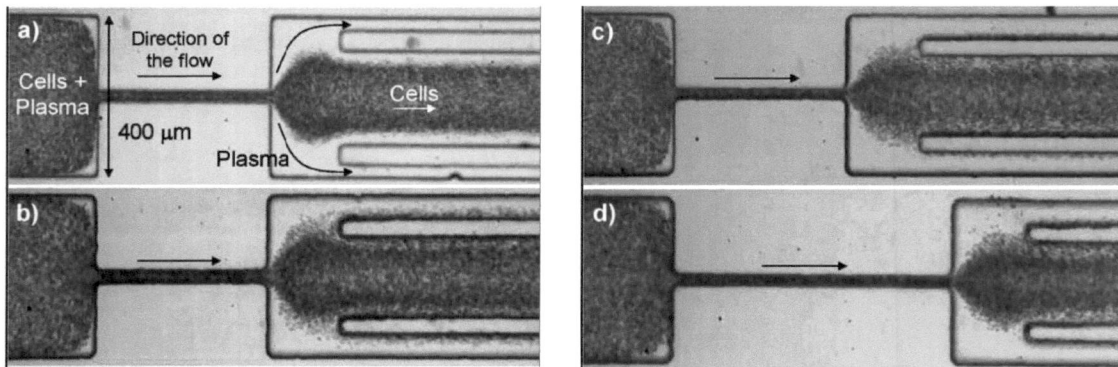

Figure 10: A microfluidic design for separating plasma from blood [79]. (**a**) $w = 25$ μm, $L = 500$ μm, $Q = 200$ μl/hr, width of the outermost channel is 30 μm. (**b**) Increase of the width of the outermost channel to 50 μm, with $w = 25$ μm, $L = 500$ μm, $Q = 200$ μl/hr. (**c**) Decrease of the constriction width to $w = 15$ μm with $L = 500$ μm, $Q = 200$ μl/hr and the width of the outermost channel is 50 μm. (**d**) Increase the length of the construction to $L = 800$ μm with $w = 25$ μm, $Q = 200$ μl/hr, and the width of the outermost channel is 50 μm. Reprinted with permission from the IOS Press. Copyright (2006)

3.6. Capillary Effect

Microfluidic capillary system was reported to autonomously transport aliquots of different liquids based on capillary phenomena [80]. The capillary effect does not require any external power supply or control device which can be used to separate plasma from whole blood.

A microdevice for blood plasma separation using capillary phenomenon was reported by Khumpuang [81] in which blood plasma was separated by microcapillary channels fabricated by silicon bulk-micromachining process. The microfluidic device consists of a silicon separation chip and a glass cover. A drop of blood was drawn into the channels and then the flow stopped in less than 2 min which indicated the completion of the separation process. Thus the separation was relatively fast (less than 2 min). After separation of blood, 1 μl of plasma was collected by using this device. Due to the advantages of small sample usage and no power consumption, the device is promising for blood diagnosis on a chip for Point-of-Care Testing (POCT).

4. CASE STUDY I: A MICROFLUIDIC DEVICE TARGETING PLASMA ISOLATION USING PLASMA SKIMMING

In this section, a specific microfluidic device developed to separate blood plasma from whole blood samples was covered in detail. Device working principle, fabrication and experimental results were included, aimed to further demonstrate the concept of applying microfluidic devices for plasma isolation [82]. This microchip enables blood separation in a continuous and real time manner with potentials not only for on-line blood biochemical tests, but also blood component transfusion and plasma exchange therapy.

4.1. Principle

Blood exhibits unique flow characteristics on the scale of the microcirculation for its particulate nature. RBCs comprising more than 99% of whole blood cells, which are deformable and at rest assume a biconcave discoid shape, tend to flow at the center of blood vessels, leaving a plasma-rich zone adjacent to the vessel wall [83].

The tendency of RBCs to concentrate at the center of the blood stream leads to plasma skimming: an uneven distribution of RBCs and plasma among the two daughter vessels of any nonsymmetrical bifurcation [84]. The vessel that receives less flow receives a disproportionately smaller fraction of RBCs and, therefore, has a lower hematocrit.

Besides the plasma skimming effect, the microcentrifugal effect was also used in this example to further improve the plasma separation efficiency. The centrifugal force (F) was given by:

$$F = m \cdot v^2/R$$

where m is the mass of the particles, v is the velocity of particles' motion and R is the radius of particles motion. Since the mass of blood cells is fixed, centrifugal forces can be increased by increasing the velocity and decreasing the radius of particles' motion. In this example, the velocity is adjusted by fluid flow pressure during the experimental process, to produce centrifugation forces for plasma and cell separation.

4.2. Design and Fabrication

The microfluidic device is shown in (Fig. **11a**). It consists of an inlet reservoir and three outlet reservoirs, connected by a network of microchannels: the entrance channel, the main semicircular channel, the sub-channels and the outflow channels. The main semicircular channel and sub-channels are connected by slots shown in (Fig. **11b**). The first outlet reservoir and outflow channel are used for blood cell collection while the second ones are used for plasma collection with the third ones used for other purposes not covered in this chapter.

The main semicircular channel is 100 μm wide to collect larger fractions of RBCs. The slots are 5 μm wide, forming the connection of the main semicircular channel with the sub-channels, which enables the collection of the pure plasma using the plasma skimming effect. The sub-channels are also named as plasma channels, which are 50μm wide, designed to collect the purified plasma from the slots.

In this design, since in the equation to calculate the centrifugal force, the radius of particles' motion is represented by the radius of the microchannels, he main semicircular channel was designed with gradually decreased radius (Fig. **11a**). This decrease in radius of the semicircular channel could further enhance the separation efficiency of plasma by increasing centrifugal forces.

The silicon substrate was fabricated by MEMS technologies. The process is as follows. A silicon wafer was first treated with oxygen plasma, and then spin coated with positive photoresist. After photolithography, microchannels, slots and reservoirs were etched by a deep reaction ion etcher. After the removal of the resist, the silicon wafer was oxidized to be biocompatible. Then the patterned silicon substrate was treated with oxygen plasma and bonded with a PDMS-glass compounded cover.

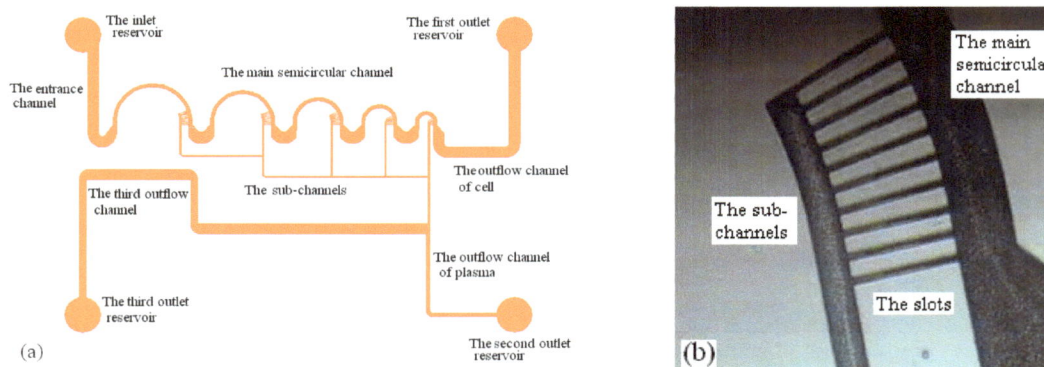

Figure 11: (a) Schematic of the microdevice for plasma isolation. **(b)** Microscopic image of the top view of the microdevice at the location of the dotted square marked in (a). These narrow slots are approximate 5μm×200μm spaced by about 35μm.

Then the bonding process between the silicon substrate and the PDMS-glass cover was shown as follows. Firstly, four holes (one as the inlet port & the others as the outlet ports) were drilled in a glass slide aligned with the features on the silicon substrate. The glass was then cleaned thoroughly. A 10:1 mixture of PDMS oligomer and the cross-linking agent (Sylgard 184), was degassed under vacuum, and then poured onto a flat silanized glass substrate. The glass with *via* holes was then gently put on to the PDMS prepolymer and cured in an oven at 80 °C for 1 hour. After the flat silanized glass slide was peeled off and the excessive PDMS in the holes was removed, the PDMS-glass cover was bonded with the patterned silicon substrate after treatment with oxygen plasma for 1 min.

Rabbit whole blood was collected from an ear into Vacutaner tubes containing EDTA-K_2. Microfluidic devices were rinsed by 20 μL 0.9% NaCl solution prior to experiments. The experimental process is summarized as follows. A diluted blood sample (anticoagulated rabbit whole blood diluted with 0.9% NaCl solution) was introduced into the microdevice at the inlet by a peristaltic pump. Then the blood sample was pushed to flow into the main semicircular channel from the entrance channel. The majority of blood cells were confined in the main semicircular channel and flushed to the outflow channel of the cell. In the meanwhile, plasma with extremely low number of cells flowed through the slots to the sub-channels, and was finally collected from the second outlet *via* the outflow channel of plasma. The blood cells in the initial whole blood samples and collected samples were counted and compared by a blood counting chamber flow cytometry.

Prior to experiments, three pipette tips were aligned and glued at the three holes of the glass cover with the fourth hole blocked up. The microfluidic device used for blood separation is shown in (Fig. **12a**). The size of this device is 1.5cm×1cm, and the inlet and outlet ports are 1mm in diameter.

Once blood plasma isolation was conducted in the microfluidic device, the performance of the device was quantified by the plasma selectivity. The plasma selectivity (σ) is defined as:

$$\sigma = 100 \ (1\text{-}C_2/C_0) \ \%$$

where C_0 and C_2 are the cell concentration in the inlet reservoir and the second outlet reservoir for plasma, respectively. A plasma selectivity of 100% indicates that no cells travel into the plasma channel and highly purified plasma is collected. (Fig. **12b**) shows the effect of cell concentration on plasma selectivity. A high plasma separation efficiency of 95.36% was achieved when the cell concentration of the blood sample was relatively low. However, the increase of cell concentration didn't lead to the decrease of the plasma selectivity significantly.

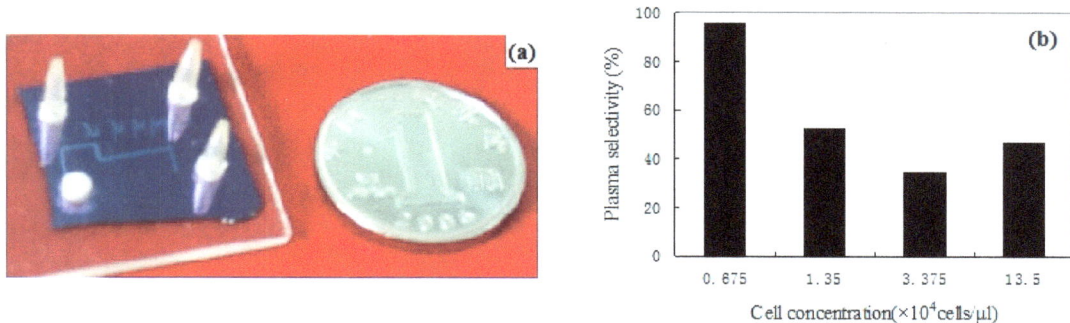

Figure 12: (**a**) Photograph of the microfluidic device. (**b**) Effect of variation of cell concentration on plasma selectivity.4.3 Experimental Results.

The effect of flow velocity on the performance of the microfluidic device was also investigated. Blood separation experiments were conducted the same microfluidic device at a variety of velocities by modulating fluid flow pressures accordingly, with other parameters unchanged. Experiment results show the effect of the flow rate ratio between the cell channel and the plasma channel on the plasma selectivity. The plasma selectivity was noted to decrease from 93.83% to 19.14% as the flow rate ratio was decreased from 13.6 to 2.3. Both the plasma skimming effect and the microcentrifugal force push RBCs to flow in the faster flow regions of the main semicircular channel, which lead to lower number of cells flowing through the narrow slots to the plasma channel. In summary, higher flow rate ratios can yield a higher plasma selectivity using the current device design.

5. CASE STUDY II: A MICROFLUIDIC CHIP FOR PLASMA ISOLATION BASED ON CROSSFLOW FILTRATION

In this example, the separation of blood plasma from whole blood is demonstrated by novel silicon nano-filters based on the crossflow filtration principle [85]. The microfluidic chip consists of a compound cover and a silicon substrate containing micropillar arrays, feed channels, side channels and nano-gap structures. The silicon structures for filtration were designed and confirmed by numerical simulations. The microfabrication procedures were investigated and optimized. The filtration structures were characterized by SEM and then used to isolate plasma from whole blood in a continuous manner. Compared with micro-gap structures in silicon microfluidic channels, the nano-gap structures were successfully used to separate plasma from the whole blood samples with higher selectivity, where the highest plasma selectivity of 97.7% was recorded. Problems of clogging or jamming were not noticed during the whole separation process. The microfluidic chip with nano-gap structures for plasma isolation could be integrated into micro total analysis system for point-of-care diagnostics in the near future.

5.1. Principle

For dead-end filtration, the fluidic flow is perpendicular to the filtration structures so that small particles pass through the filtration barriers along the fluid, while larger particles are stopped to cause channel clogging or jamming. To overcome clogging of dead-end filtration, crossflow filtration, which allows the bigger particles to stay in a suspended state instead of being aggregated after being stopped, have been used in microfluidic chips.

For the crossflow filtration, the flow direction is parallel to the filtration structures. The tangential rate is used for transporting the permeated fluid from the feed stream to the lateral stream through the filtration structures, and the retentive particles are still flowing along the feed stream. This method can dramatically alleviate the problem of clogging or jamming, which is the key feature of this example, shown as follows.

5.2. Design and Simulation

In this case study, several filtering structures were proposed *via* the crossflow filtration principle, and then compared by numerical simulation results using FLUENT (Version: 2d, double precision, segregated, laminar, Release 6.2.16). As shown in (Fig. **13**), the design with two inlets is capable of transferring more fluid from the feed stream to the side stream than the design with one inlet. In the design with two inlets shown in (Fig. **13b**), the second inlet "2" perpendicular to the filtration structures was used to obtain higher tangential rates. In order to drive fluid flow using only one pump, the two inlets are connected by microchannels, which showed comparable performance with the models with two separate inlets, verified by numerical simulations.

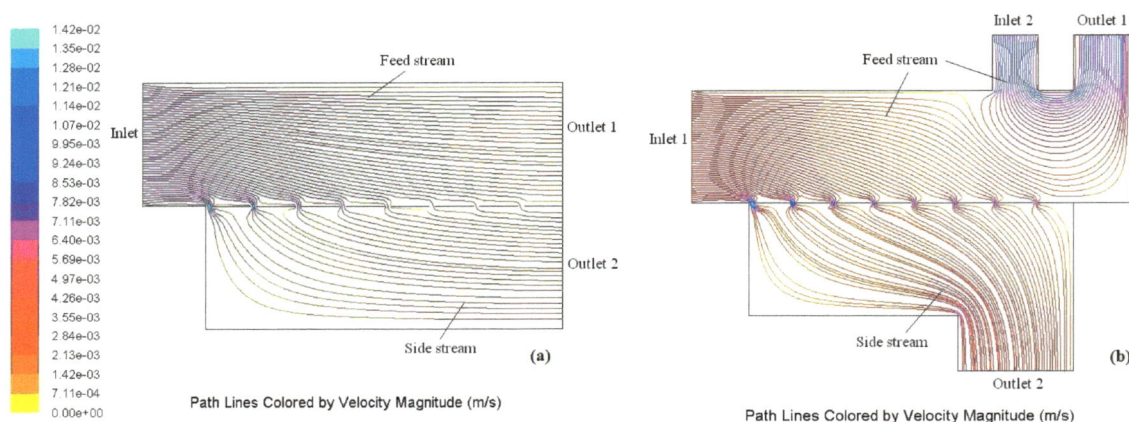

Figure 13: Numerical simulation results for the streamline distribution at different models. (**a**): Design model with one inlet; (**b**): Design model with two inlets. The path lines represent streamlines of the liquid.

Based on the simulation results, the optimized design of the microfluidic structure was fabricated by MEMS techniques. In the optimized design, the silicon substrate mainly includes microchannels, square micropillars and reservoirs, shown in (Fig. **14**). The bypass channels that are perpendicular to the filtration structures, shown in (Fig. **14**), are used as "inlets" to increase tangential flow rates. The square micropillars with nano-gaps divide the whole microchannel into a feed microchannel and a side microchannel. In addition, the serpentine structure in the waste outlet channel is used to increase the flow resistance, which can also enhance the tangential flow rate.

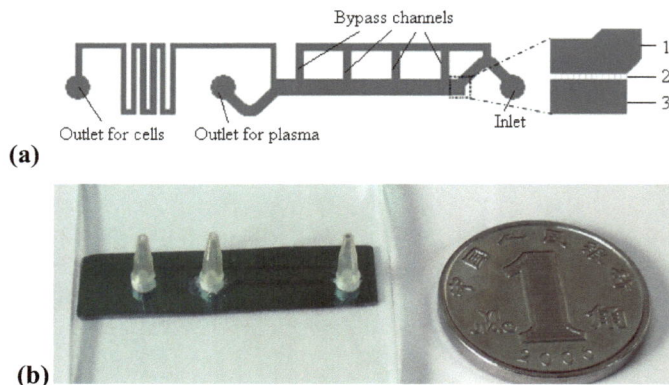

Figure 14: Layout of microfluidic chip for plasma isolation. (**a**): Schematic of the microchip, "1" the feed microchannel, "2" square micropillars array, "3" the side microchannel; (**b**): photograph of the microchip.

A prototype of the fabricated microfluidic chip consisting of a PDMS-glass compound cover and a silicon substrate is shown in (Fig. **14b**), with the dimensions of 2 cm × 0.5 cm.

5.3. Fabrication, Characterization and Geometry Optimization of Micropillars with Nanogaps

It is well known that RBCs comprising more than 99% of whole blood cells at rest assume a biconcave discoid shape with a diameter of ~8 μm and a thickness of ~2 μm and they are capable of passing through capillaries with less than half of their dimensions. Removing RBCs is the key point for plasma isolation and is the major purpose of the proposed microfluidic separation chip. In this chip, the width of the filtration structures was designed to be less than 1 μm to let pass plasma and filter out red blood cells.

In order to obtain nano-gap structures and deal with the challenges of the MEMS technologies, the fabrication procedures and relevant parameters were carefully investigated by experiments. A Cr photolithography mask in which the 1 μm gaps between large square micropillars of 50 μm × 50 μm was designed and fabricated by the Electron-beam Lithography.

Initially, the conventional MEMS procedures including a photolithograph step, an etching etch and a surface treatment step were used, and the obtained gap structures were characterized by SEM. The size of gaps fabricated with the conventional fabrication process in our lab was around 3.6 μm, which is much higher than the dimension of the original design and therefore cannot be used for plasma separation.

Since the photolithograph is the first important step in the whole fabrication procedure, including three sub-steps: spin coating, exposure and development, the parameters for each sub step were optimized individually. The silicon wafer was spin coated with positive photoresist at higher speeds to obtain thinner photoresist films, which should lead to higher geometry resolution. The thickness of the films obtained at the speeds of 1000~1500 rpm/min, 2000~3000 rpm/min and 3000~4000 rpm/min were 2.1~3μm, 1.4~1.8μm and 1.1~1.4μm, respectively. For the thinner photoresist film, the exposure time was reduced accordingly in order to avoid overexposing which could result in broadening of the patterned gaps. At the same time the development time was decreased and the concentration of NaOH buffer for developing was also decreased to deal with the issue of overdevelopment.

After several groups of control experiments, the optimal parameters in photolithography were briefly summed up as follows. The speed of spinning was 3000~4000 rpm/min which was the up limitation of our spin coater, whereas the exposure time was 11~13 sec. The concentration of NaOH buffer for development was 0.6% by weight and the development time was 30~40 sec. The size of the gaps fabricated with the optimal photolithography parameters was about 2.1 μm, which was closer to the dimension of the original design.

Since the obtained ~ 2 μm micro-gaps cannot meet the requirement of plasma isolation when the deformability of red blood cells is considered, the fabrication procedures were optimized further. Initially, thermally oxidization for an hour was conducted on patterned silicon wafer to make the surface biocompatible. With a longer oxidization time, the SiO_2 film is gradually thickened, enabling the fabrication of gaps in the nano-scale, which can be used to deal with the deformability of RBCs. Thus a special 7 hour thermal oxidization procedure was used in this study to decrease the gap further. The obtained gap structures were also characterized by SEM, which demonstrates that the narrow slits of approximate 800 nm were obtained between the square microposts of 50μm×50μm by the special thermal oxidization for 7 hours (see Fig. **15a** and Fig. **15b**). It is worth noting that boarder slits of about 1.9 μm were observed after 1 hour oxidation in (Fig. **15c** and **15d**), which further indicates that longer oxidation time does help decrease the gap of micropillars significantly.

5.4. Plasma Separation Experiments

Experiments of blood separation were conducted by using two types of microfluidic chips with micro and nano-scale gaps respectively. Prior to experiments, 20 μL 0.9% NaCl solution was introduced into the microfluidic chip for rinsing channel by a peristaltic pump. In a typical experiment of plasma separation, the diluted blood sample was introduced into the chip at the inlet. The blood cells in the initial whole blood sample and collected samples from outlets were counted by a blood counting chamber flow cytometry.

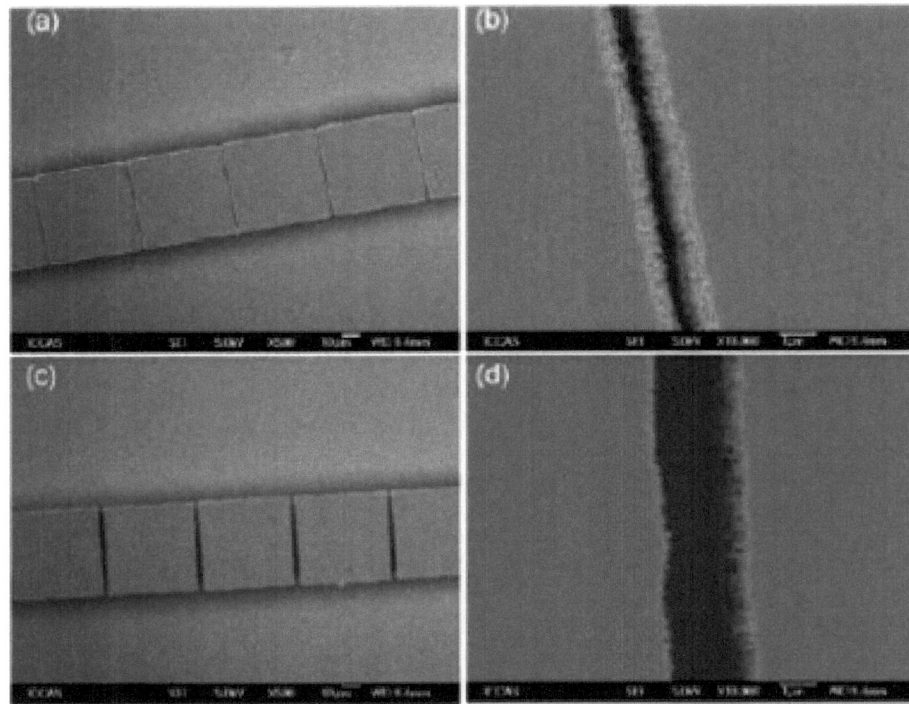

Figure 15: SEM of nano-slit structure between 50μm×50μm square microposts by nomal thermal oxidization for 7h (**a** and **b**) and by special thermal oxidization for 1h (**c** and **d**). (a), (c) ×500; (b), (d) ×10,000.

The first microfluidic chip with micro-gap structures of about 2 μm in width was used to separate plasma from whole blood. The process of blood separation was monitored by a microscope with a CCD camera and a video monitoring system. As is shown in (Fig. **16a**), a large number of blood cells were noted to enter the side microchannel through the micro-gaps from the feed microchannel, resulting into the failure of plasma isolation. These results agreed well with values reported by Sethu [86] who reported a microfluidic device where 50%~75% of RBCs and 3%~17% WBCs passed through the micro-structures and flowing out at outlets 1 and 3.

The second microfluidic chip with nano-gap structures was used to separate whole blood samples as a comparison. The experimental conditions of the second chip were the same as that of the first one except that two whole blood samples with different cell concentrations were used and compared. The processes of blood separation were also monitored by a microscope with a CCD camera and a video monitoring system. The separation process is shown in (Fig. **16b**) which indicated that as a small quantity of plasma passed through the nano-gap structures, only a few blood cells passed through the nano-filters and the majority of blood cells still remained in the feed channel. For the blood sample with higher cell concentration, plasma passed through the nano-gaps quickly initially and then the rate of plasma entering the side microchannel reduced sharply due to the accumulation of blood cells around nano-gap structures. For the blood sample with a lower cell concentration, almost all the blood cells remained in the feed channel and only plasma passed through the nano-filters. Compared to the case with a higher cell concentration, the rate of plasma passing through the nano-gap was much higher and the separated plasma flowed out much faster. Images and videos indicates that the nano-gap structures in the microfluidic chip were successfully used to separate plasma from whole blood samples, while most blood cells were prevented from entering the side microchannel.

The images and videos could only qualitatively prove that the nano-gap structures in the microfluidic chip enables effective isolatation of plasma from whole blood compared with the micro-gap structures. After blood plasma isolation was completed in the proposed microchip, the performance of the nano-gap structures in the microfluidic chip was quantified by the plasma selectivity (σ) which is defined as:

$$\sigma = 100(1\text{-}C_2/C_0) \ \%$$

where C_0 and C_2 are the cell concentration in the inlet reservoir and the second outlet reservoir for plasma, respectively. A plasma selectivity of 100% indicated that no cells travel into the plasma channel.

Blood samples with different cell concentrations were used to evaluate the performance of the chip. The plasma selectivity increased from 44.7% to 97.7% when the cell concentration of the blood samples used was reduced from 3.67×10^4 to 2.8×10^3 per microliter. With other parameters unchanged, these experimental results show that the diluted blood samples produced higher selectivity compared to undiluted samples.

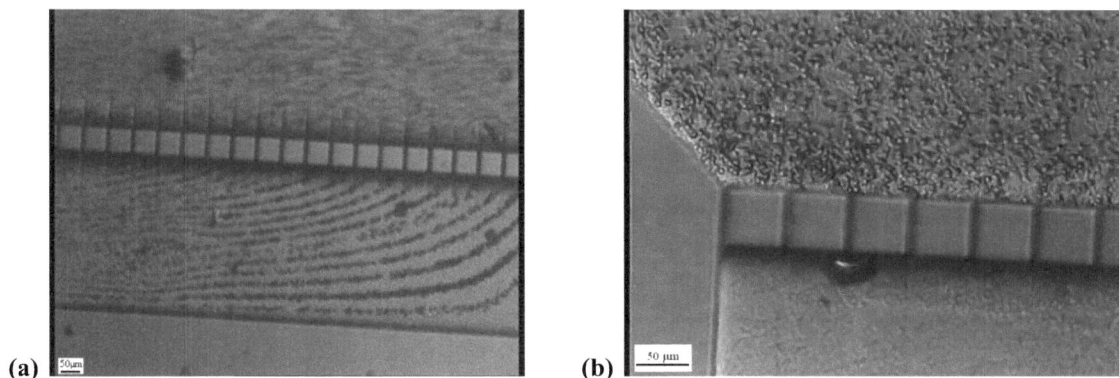

Figure 16: (**a**) Image of plasma separation by using the first chip with micro-gap structures. (**b**) Image of isolation plasma from whole blood with higher cell concentration by using the second chip with nano-gap structures.

The pressure is another factor that affects blood separation efficiency through nano-filters. For the experiments of blood separation, the blood fluid was driven by a peristaltic pump while the outlets were free outflow boundary conditions so that the pressure related to flow rate was controlled by the pump. For a fluid with viscosity (η) and flow rate (Q), the differential pressure (ΔP) through a channel of width (w), depth (h), and length (l) is given by $\Delta P = Ql\eta/kwh^3$, where k is a constant that depends on the ratio w/h [87]. The higher the flow rate, the higher differential pressure needed under the same other conditions. In the experiment, when the flow rate was too high, (*i.e.*, 25 µl/min at inlet), blood cell lysis occurred due to high pressure. The ruptured cells contaminated the isolated plasma, while the cell debris jammed the nano-filters resulting in separation failure. When the flow rate at inlet was too low (*e.g.*, 1 µl/min), the differential pressure was too low to push plasma through nano-filters and resulted in limited collection of plasma at the plasma outlet. Therefore the flow rate at inlets should be controlled in a range from 8 µl/min to 15 µl/min. Under the optimal flow rate, the velocity of plasma isolation was 7.5 µl/min in the side channels with the cell concentration of 2.8×10^3 per microliter.

Continuous blood separation experiments were conducted for more than one hour without any problems of jamming or clogging under the optimal conditions. The plasma and the blood cells flowing out from each outlet were collected respectively, which could be used for further clinic assays.

5.5. CONCLUSION

In this example, MEMS technologies were used to fabricate nano-gap structures in a microfluidic chip by optimizing microfabrication procedures. Compared with the micro-gap structures, the nano-gap structures presented in this example was capable of isolating plasma isolation from whole blood samples in a continuous, real-time manner without problems of jamming or clogging. Plasma and blood cells were collected individually for subsequent analysis. This microchip for size-dependent separation has been shown to exhibit several distinct advantages over its large-scale counterparts, including low blood consumption, high speed, high separation efficiency and portability. This device could be further integrated with other microfluidic devices to form µTAS platforms for blood clinic assays.

ACKNOWLEDGEMENT

The authors greatly acknowledge the financial support from the National Science Foundation of China under the Grant numbers of 60701019 and 60501020.

REFERENCES

[1]   Lanzer G, Tilz GP. Therapeutic separation of blood components in clinical emergency medicine. Wien Med Wochenschr 1986; 136(5-6): 134-7.

[2]   Malchesky PS, Wojcicki J, Nose Y. Advances in blood component separation and plasma treatment for therapeutics. J Parenter Sci Technol 1983; 37(1): 2-4.

[3]   Vettore L, De Matteis MC, Zampini P. A new density gradient system for the separation of human red blood cells. Am J Hematol 1980; 8(3): 291-7.

[4]   Zizzadoro C, Belloli C, Badino P, Ormas P. A rapid and simple method for the separation of pure lymphocytes from horse blood. Vet Immunol Immunopathol 2002; 89(1-2): 99-104.

[5]   Toth TE, Smith B, Pyle H. Simultaneous separation and purification of mononuclear and polymorphonuclear cells from the peripheral blood of cats. J Virol Methods 1992; 36(2): 185-95.

[6]   Targowski SP. Separation of mononuclear leukocytes and polymorphonuclear leukocytes from equine blood. Can J Comp Med 1976; 40(3): 285-90.

[7]   Buckner D, Eisel R, Perry S. Blood cell separation in the dog by continuous flow centrifugation. Blood 1968; 31(5): 653-72.

[8]   AuBuchon JP, Dumont LJ, Herschel L, *et al.* Automated collection of double red blood cell units with a variable-volume separation chamber. Transfusion 2008; 48(1): 147-52.

[9]   Stromberg RR, Friedman LI, Boggs DR, Lysaght MJ. Membrane technology applied to donor plasmapheresis. J Memb Sci 1989; 44(1): 131-43.

[10]   Ding LH, Jaffrin MY, Gupta BB. A model of hemolysis in membrane plasmapheresis. ASAIO transactions / American Society for Artificial Internal Organs 1986; 32(1): 330-3.

[11]   Friedman LI, Hardwick RA, Daniels JR. Evaluation of membranes for plasmapheresis. Artif Organs 1983; 7(4): 435-42.

[12]   Malchesky PS, Horiuchi T, Lewandowski JJ, Nose Y. Membrane plasma separation and the on-line treatment of plasma by membranes. J Memb Sci 1989; 44(1): 55-88.

[13]   Solomon BA, Castino F, Lysaght MJ. Continuous flow membrane filtration of plasma from whole blood. Transactions of the American Society for Artificial Organs. 1978;VOL.24:21-6.

[14]   Zydney AL, Colton CK. Continuous flow membrane plasmapheresis: theoretical models for flux and hemolysis prediction. Transactions - American Society for Artificial Internal Organs; 1982; 28: 408-12.

[15]   Onishi M, Shimura K, Seita Y, Yamashita S, Takahashi A, Masuoka T. Preparation and properties of plasma-initiated graft copolymerized membranes for blood plasma separation. Int J Rad Appl Instrum A. 1992; 39(6): 569-76.

[16]   Natori SH, Kurita K. Blood cell separation using crosslinkable copolymers containing N,N-dimethylacrylamide. Polym Adv Technol 2007; 18(4): 263-7.

[17]   Higuchi A, Yang ST, Li PT, *et al.* Separation of hematopoietic stem and progenitor cells from human peripheral blood through polyurethane foaming membranes modified with several amino acids. J Appl Polym Sci 2009; 114(2): 671-9.

[18]   Higuchi A, Sekiya M, Gomei Y, *et al.* Separation of hematopoietic stem cells from human peripheral blood through modified polyurethane foaming membranes. J Biomed Mater Res A 2008; 85(4): 853-61.

[19]   Zhou D, Stewart GJ, Kowalska MA, Niewiarowski S. An improved method for separation of neutrophils from human blood using methylcellulose. Thromb Res 1995; 80(3): 271-5.

[20]   Higuchi A, Yamamiya SI, Yoon BO, Sakurai M, Hara M. Peripheral blood cell separation through surface-modified polyurethane membranes. J Biomed Mater Res A 2004; 68(1): 34-42.

[21]   Brody J, Osborn T, Forster F, Yager P. A planar microfabricated fluid filter. Sens Actuators A Phys 1996; 54(1-3): 704-8.

[22]   Tachi T, Kaji N, Tokeshi M, Baba Y. Simultaneous separation, metering, and dilution of plasma from human whole blood in a microfluidic system. Anal Chem 2009; 81(8): 3194-8.

[23] Vezon G, Piquet Y, Manier C, Schooneman F, Mesnier F, Moulinier J. Technical aspects of different donor plasmapheresis systems and biological results obtained in collected plasma. Vox Sang 1986; 51(Suppl 1): 40-4.

[24] Piquet Y, Vezon G, Schooneman F, *et al.* Plasmapheresis in normal donors: comparative study of two methodologies: centrifugation and filtration. Plasma Ther Transfus Technol 1985; 6: 415-20.

[25] Rock G, Tittley P, McCombie N. Plasma collection using an automated membrane device. Transfusion 1986; 26(3): 269-71.

[26] Perseghin P, Pagani A, Fornasari P, Salvaneschi L. Donor plasmapheresis: a comparative study using four different types of equipment. Int J Artif Organs 1987; 10(1): 51.

[27] Gurland H, Lysaght M, Samtleben W, Schmidt B. Comparative evaluation of filters used in membrane plasmapheresis. Nephron 1984; 36(3): 173-82.

[28] Malbrancq J, Jaffrin M, Bouveret E, Angleraud R, Vantard G. Plasmafiltration through a plane microporous membrane. ASAIO J 1984; 7(1): 16-24.

[29] Friedman L, Hardwick R, Daniels J, Stromberg R, Ciarkowski A. Evaluation of membranes for plasmapheresis. Artif Organs 1983; 7: 435-42.

[30] Buffaloe G, Erickson R, Dau P. Evaluation of a parallel plate membrane plasma exchange system. J Clin Apheresis 1983; 1: 86-94.

[31] Sato H, Malchesky P, Nos¨¦ Y. Characterization of polypropylene membrane for plasma separation. Artif Organs 2008; 7(4): 428-34.

[32] Stromberg R, Lysaght M, Boggs D, *et al.* Development of a novel membrane apheresis system for plasma collection at mobile sites. ASAIO J 1987; 33(3): 614.

[33] Lysaght M, Samtleben W, Schmidt B, Stoffner D, Gurland H. Spontaneous membrane plasmapheresis. Transactions-American Society for Artificial Internal Organs 1983; 29:506.

[34] Schmidt B, Lysaght M, Samtleben W, Gurland H. Plasmapheresis without pumps for therapeutic and donor purposes. Plasma Separation and Plasma Fractionation: Current Status and Future Directions 1983: 188.

[35] Landini S, Coli U, Lucatello S, Fracasso A, Morachiello P, Righetto F, *et al.* Spontaneous plasma exchange by gravity. Int J Artif Organs 1984; 7(3): 137.

[36] Ji HM, Samper V, Chen Y, Heng CK, Lim TM, Yobas L. Silicon-based microfilters for whole blood cell separation. Biomed Microdevices. 2008; 10(2): 251-7.

[37] Wilding P, Pfahler J, Bau HH, Zemel JN, Kricka LJ. Manipulation and flow of biological fluids in straight channels micromachined *in silicon*. Clin Chem1994; 40(1): 43-7.

[38] Wilding P, Kricka LJ, Cheng J, Hvichia G, Shoffner MA, Fortina P. Integrated cell isolation and polymerase chain reaction analysis using silicon microfilter chambers. Anal Biochem 1998; 257(2): 95-100.

[39] He B, Tan L, Regnier F. Microfabricated filters for microfluidic analytical systems. Anal Chem 1999 Apr 1; 71(7): 1464-8.

[40] Lettieri G, Dodge A, Boer G, Rooij N, Verpoorte E. A novel microfluidic concept for bioanalysis using freely moving beads trapped in recirculating flows. Lab Chip 2003; 3(1): 34-9.

[41] Svec F, Peters E, Sykora D, Yu C, Frechet J. Monolithic stationary phases for capillary electrochromatography based on synthetic polymers: designs and applications. J High Resolut Chromatogr 2000; 23(1).

[42] Crowley TA, Pizziconi V. Isolation of plasma from whole blood using planar microfilters for lab-on-a-chip applications. Lab Chip 2005; 5(9): 922-9.

[43] Moorthy J, Beebe D. *In situ* fabricated porous filters for microsystems. Lab Chip 2003; 3(2): 62-6.

[44] Thorslund S, Klett O, Nikolajeff F, Markides K, Bergquist J. A hybrid poly(dimethylsiloxane) microsystem for on-chip whole blood filtration optimized for steroid screening. Biomed Microdevices 2006; 8(1): 73-9.

[45] Sollier E, Rostaing H, Pouteau P, Fouillet Y, Achard JL. Passive microfluidic devices for plasma extraction from whole human blood. Sensor Actuator B: Chem 2009; 141(2): 617-24.

[46] VanDelinder V, Groisman A. Separation of plasma from whole human blood in a continuous cross-flow in a molded microfluidic device. Anal Chem 2006 ; 78(11): 3765-71.

[47] Schembri C, Burd T, Kopf-Sill A, Shea L, Braynin B. Centrifugation and capillarity integrated into a multiple analyte whole blood analyser. J Automat Chem 1995; 17: 99-104.

[48] Steigert J, Grumann M, Brenner T, *et al.* Integrated sample preparation, reaction, and detection on a high-frequency centrifugal microfluidic platform. Journal of the Association for Laboratory Automation 2005; 10(5): 331-41.

[49] Grumann M, Geipel A, Riegger L, Zengerle R, Ducr¨¦e J. Batch-mode mixing on centrifugal microfluidic platforms. Lab Chip 2005; 5(5): 560-5.

[50] Lai S, Wang S, Luo J, Lee L, Yang S, Madous M. Design of a compact disk-like microfluidic platform for enzyme-linked immunosorbent assay. Anal Chem 2004; 76(7): 1832-7.

[51] Puckett L, Dikici E, Lai S, Madou M, Bachas L, Daunert S. Investigation into the Applicability of the Centrifugal Microfluidics Platform for the Development of Protein Ligand Binding Assays Incorporating Enhanced Green Fluorescent Protein as a Fluorescent Reporter. Anal Chem 2004; 76(24): 7263-8.

[52] Haeberle S, Brenner T, Zengerle R, Ducrée J. Centrifugal extraction of plasma from whole blood on a rotating disk. Lab Chip - Miniaturisation Chem Biol 2006; 6(6): 776-81.

[53] Zhang J, Guo Q, Liu M, Yang J. A lab-on-CD prototype for high-speed blood separation. J Micromech Microeng 2008;18(12).

[54] Steigert J, Brenner T, Grumann M, *et al.* Integrated siphon-based metering and sedimentation of whole blood on a hydrophilic lab-on-a-disk. Biomed Microdevices 2007; 9(5): 675-9.

[55] Lee BS, Lee JN, Park JM, Lee JG, Kim S, Cho YK, *et al.* A fully automated immunoassay from whole blood on a disc. Lab Chip 2009; 9(11): 1548-55.

[56] Yosioka K, Kawasima Y. Acoustic radiation pressure on a compressible sphere. Acustica 1955; 5(3):167-73.

[57] Cousins C, Holownia P, Hawkes J, *et al.* Plasma preparation from whole blood using ultrasound. Ultrasound Med Biol 2000; 26(5): 881.

[58] Petersson F, Nilsson A, Holm C, Jönsson H, Laurell T. Continuous separation of lipid particles from erythrocytes by means of laminar flow and acoustic standing wave forces. Lab Chip - Miniaturisation Chem Biol 2005; 5(1): 20-2.

[59] Petersson F, Nilsson A, Holm C, Jönsson H, Laurell T. Separation of lipids from blood utilizing ultrasonic standing waves in microfluidic channels. Analyst 2004; 129(10): 938-43.

[60] Petersson F, Nilsson A, Holm C, Jönsson H, Laurell T. Continuous separation of lipid particles from erythrocytes by means of laminar flow and acoustic standing wave forces. Lab Chip 2005; 5(1): 20-2.

[61] Petersson F, Nilsson A, Jönsson H, Laurell T. Carrier medium exchange through ultrasonic particle switching in microfluidic channels. Anal Chem 2005; 77(5): 1216-21.

[62] Lenshof A, Ahmad-Tajudin A, Jaras K, *et al.* Acoustic whole blood plasmapheresis chip for prostate specific antigen microarray diagnostics. Anal Chem 2009; 81(15): 6030-7.

[63] Nilsson A, Petersson P, Laurell T. Whole blood plasmapheresis using acoustic separation chips. Proceedings of the Micrototal Analysis Systems 2006: 314-6.

[64] Nilsson A, Petersson F, Jönsson H, Laurell T. Acoustic control of suspended particles in micro fluidic chips. Lab Chip - Miniaturisation Chem Biol 2004; 4(2): 131-5.

[65] Yang S, Zahn JD. Particle separation in microfluidic channels using flow rate control. ASME International Mechanical Engineering Congress, Anaheim, California, USA 2004.

[66] Yang S, Ündar A, Zahn JD. A microfluidic device for continuous, real time blood plasma separation. Lab on a Chip - Miniaturisation Chem Biol 2006; 6(7): 871-80.

[67] Yang S, Ündar A, Zahn JD. Blood plasma separation in microfluidic channels using flow rate control. ASAIO J 2005; 51(5): 585-90.

[68] Yang S, Ündar A, Zahn JD. Biological fluid separation in microfluidic channels using flow rate control. ASME International Mechanical Engineering Congress and Exposition, Orlando, Florida, USA 2005.

[69] Yang S, Ji B, Undar A, Zahn JD. Microfluidic devices for continuous blood plasma separation and analysis during pediatric cardiopulmonary bypass procedures. ASAIO J 2006; 52(6): 698-704.

[70] Kersaudy-Kerhoas M, Dhariwal R, Desmulliez MPY, Jouvet L. Hydrodynamic blood plasma separation in microfluidic channels. Microfluid Nanofluidics 2009: 1-10.

[71] Kersaudy-Kerhoas M, Dhariwal R, Desmulliez M, Jouvet L. Blood flow separation in microfluidic channels. Proc μFlu'08. 2008.

[72] Park J, Cho K, Chung C, Han DC, Chang JK. Continuous plasma separation form whole blood using microchannel geometry. 3rd IEEE/EMBS Special Topic Conference on Microtechnology in Medicine and Biology, Kahuku, Oahu, USA 2005.

[73] Barbee J, Cokelet G. The Fahraeus effect. Microvasc Res 1971; 3(1): 6.

[74] Fischer T, St hr-Lissen M, Schmid-Sch nbein H. The red cell as a fluid droplet: tank tread-like motion of the human erythrocyte membrane in shear flow. Science 1978; 202(4370): 894.

[75] Goldsmith H. Red cell motions and wall interactions in tube flow. Fed Proc 1971; 30(5): 1578-90

[76] Goldsmith H, Marlow J. Flow behaviour of erythrocytes. I. Rotation and deformation in dilute suspensions. Proceedings of the Royal Society of London Series B, Biological Sciences 1972; 182(1068): 351-84.

[77] Goldsmith H, Cokelet G, Gaehtgens P. Robin Fahraeus: evolution of his concepts in cardiovascular physiology. Am J Physiol 1989; 257(3): H1005.

[78] Pries A, Ley K, Gaehtgens P. Generalization of the fahraeus principle for microvessel networks. Am J Physiol 1986; 251(6): H1324.

[79] Faivre M, Abkarian M, Bickraj K, Stone HA. Geometrical focusing of cells in a microfluidic device: An approach to separate blood plasma. Biorheology 2006; 43(2): 147-59.

[80] Juncker D, Schmid H, Drechsler U, *et al.* Autonomous microfluidic capillary system. Anal Chem 2002; 74(24): 6139-44.

[81] Khumpuang S, Tanaka T, Aita E, *et al.* Blood plasma separation device using capillary phenomenon. TRANSDUCERS and EUROSENSORS '07 - 4th International Conference on Solid-State Sensors, Actuators and Microsystems, Lyon, France 2007.

[82] Chen X, Cui D, Zhang L. Microdevice for continuous isolation of plasma from whole blood. 1st International Conference on BioMedical Engineering and Informatics, Sanya, China 2008.

[83] Goldsmith H, Marlow J. Flow behavior of erythrocytes. J Colloid Interface Sci 1979; 71: 383-407.

[84] Palmer A. Influence of absolute flow rate and rouleau formation on plasma skimming *in vitro*. Am J Physio 1969; 217: 1339-45.

[85] Chen X, Cui DF, Zhang LL. Isolation of plasma from whole blood using a microfludic chip in a continuous cross-flow. Chin Sci Bull 2009; 54(2): 324-7.

[86] Sethu P, Sin A, Toner M. Microfluidic diffusive filter for apheresis (leukapheresis). Lab Chip 2006; 6(1): 83-9.

[87] VanDelinder V, Groisman A. Perfusion in microfluidic cross-flow: Separation of white blood cells from whole blood and exchange of medium in a continuous flow. Anal Chem 2007; 79(5): 2023-30.

CHAPTER 3

Microfluidic Chips for Blood Cell Separation

Xing Chen[*], Dafu Cui and Jian Chen

State Key Laboratory of Transducer Technology, Institute of Electronics, Chinese Academy of Sciences, Beijing, China

Abstract: Various methods have been demonstrated in literature for blood cell sorting and separation as one essential step of blood sample pretreatment in both the macro and micro scale. In this chapter, the latest development of the cell/particle separation using microfluidic devices is first reviewed. Furthermore, two specific microfluidic chips developed in our group targeting blood cell separation are discussed in detail as case studies. In these two examples, pillar-type and weir-type filtration structures were designed and fabricated by microelectromechanical system (MEMS) technologies, aimed to remove Red Blood Cells (RBCs) from White Blood Cells (WBCs) based on their size differences. In these two microfluidic chips, the effects of relevant parameters (*e.g.,* cell concentration variations and the dimensions of separation channels) on cell separation efficiency were investigated. Under the optimal condition, more than 95% RBC can be removed from the initial whole blood, while 27.4% WBC can be obtained.

Keywords: Blood cell separation, crossflow filtration, MEMS, microfluidics.

1. INTRODUCTION

It is well known that peripheral blood consists of RBCs that account for ~99% of the total cells and WBCs make up less than 1% of blood. In order to obtain accurate information in the downstream analytical steps, it is strongly necessary to sort, trap or concentrate the cells or virus of interest from complex whole blood samples. As an example of genomic analysis, the large population of RBCs needs to be removed from the nucleated WBCs aiming to remove the interfering substances.

There are roughly 5×10^9 cells in each milliliter of blood, including RBCs, reticulocytes, platelets, and WBCs, which are most often the target of analysis. For every leukocyte in blood there are a thousand more RBCs with it. If WBCs are the target, all RBCs need to be removed, as they do not carry useful information and may even interfere with the subsequent analysis of the leukocytes. Additionally, leukocytes themselves are also very diverse too. As we know, the classical separation of WBCs into five classes, neutrophils, lymphocytes, monocytes, eosinophils, and basophils, is already too general for many applications.

For another example, if there are 100 copies HIV in 1ml blood sample which also consisted of 5×10^9 RBCs and $5 \sim 10 \times 10^6$ WBCs, it is almost impossible to directly detect HIV before the few virus particles are trapped and enriched from the large population of blood cells. Microfabricated technologies make it possible to isolate a few cells (or even single cells) from a large population of cells on the basis of physical and chemical properties.

In this chapter, we first introduce the traditional methods for blood separation, and then review the reported microfluidic chips capable of blood isolation including the separation of RBCs from whole blood samples, the sorting of WBCs from whole blood, trapping and extracting specific cells (*e.g.,* rare tumor cells) or bacteria (*e.g., E. coli*) from whole blood samples.

In addition, two microfluidic devices developed in our group are discussed as research cases, aimed to illuminate how to design and fabricate these microfluidic devices for blood cell separation. In these two examples, the design, simulation and fabrication of these microfluidic channels are covered in detail. The

---

*Address correspondence to Xing Chen: State Key Laboratory of Transducer Technology, Institute of Electronics, Chinese Academy of Sciences, Beijing 100190, P.R. China; Tel and Fax: +86-10-58887188; E-mail: xchen@mail.ie.ac.cn

microfluidic chips for implementing blood separation in a continuous, real time manner have great potentials not only for blood biochemical tests on-line, but also for blood component transfusion and therapy.

2. TRADITIONAL METHODS FOR BLOOD CELL SEPARATION

Rapid cell sorting is very important in the newly developing fields of clinic diagnosis, cellular therapy and in biotechnology.

Density-gradient centrifugation has been the most commonly used technique for blood cell separation on the basis of differing intrinsic densities. This method has been performed utilizing many different substances, such as bovine serum albumin [1, 2], gum acacia [3], phthalate esters [4], Ficoll [5], Ludox with polyvinylpyrrolidone [6], dextran [7], arabino-galactan [8], Renografin with Ficoll [9], Renografin with arabino-galactan [10] and colloidal silica particles coated with polyvinylpyrrolidone [11].

Sedimentation centrifugation is also used to separate blood cells due to their different "effective" sizes, or, roughly, difference in molecular weights [12]. This is a "dynamic" situation because the experimental objects have started at the top and are settling through the gradient at rates roughly proportionate to the square roots of their molecular weights.

Furthermore, some traditional clinical devices rely on immunoaffinity columns, or high-gradient magnetic separation columns for blood cell separation by using either micrometer-sized polymeric beads doped with magnetite, or nanometer-size iron-dextran colloids, conjugated to targeting antibodies [13, 15].

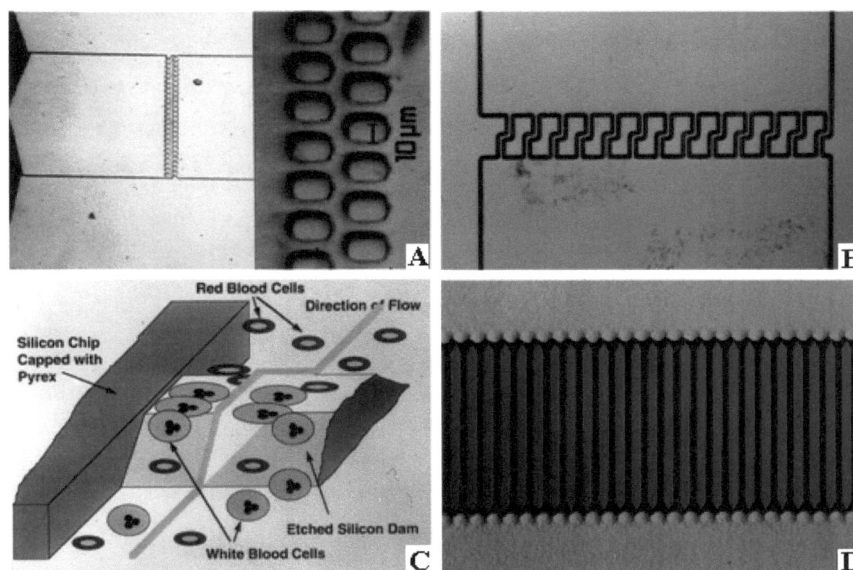

Figure 1: Silicon micropost-type and weir-type filter designs and filter chips [16]. (**A**) Offset array of simple microposts (13×20μm spaced 7μm apart). (**B**) Array of complex microposts (73μm wide) separated by 7μm wide tortuous channels spaced 30μm apart. (**C**) A 3.5μm gap between the top of the etched silicon dam and the Pyrex glass cover. (**D**) Comb-type filter formed from an array of 120 posts (175μm ×18μm spaced 6μm apart). Reprinted with permission from Elsevier. Copyright (1998).

A fast, continuous process of cell separation based on monoclonal antibodies, magnetic colloid, and a flow field in an open-gradient magnetic field was developed lately [17]. The magnetic force acting on magnetically labeled cells in such a field has a "centrifugal" character which provides a basis for the design of a continuous separation process. The continuous separation process offers advantages such as the ability to process large cell volumes, the use of on-line monitoring of the separation process, and the ability to stage and recirculate sorted fractions. However it is difficult to exert accurate control over forces involved in the displacement of the labeled cell population from the unlabeled population to a degree dictated by the resolution limitation of this method.

Based on the measured magnetic moments of hemoglobin and its compounds, and the relatively high hemoglobin concentration of human erythrocytes, magnetophoresis has been used as an approach to characterize and separate cells for biochemical analysis [18]. For example, novel particles have been fabricated for the direct magnetic separation of immune cells from whole blood based on magnetophoresis [19].

3. CURRENT MICRODEVICES TARGETING BLOOD CELL FILTRATING, SORTING AND COLLECTION

With the development of lab-on-chip technologies, chemical and biological experiments can be performed on all-in-one, automatic lab-on-chip devices, which require less reagent, smaller sample volume and a shorter assay turnaround time. Among these lab-on-chip devices, on-chip blood separation of whole blood samples is under focus, pioneered by several groups using a variety of principles.

3.1. Microstructure-Based Filtration

Filtration is an efficient method for solid particle-liquid separation, which is also used for separating particles/cells based on their size differences. The particles/cells are selective segregated by passing through filtering membranes or pillar arrays.

Wilding [20] reported a glass-silicon microfluidic chip with microchannels and micromachined filters. Microchannels and microfilters were fabricated in silicon wafer by using planar photolithography and reactive ion etching technologies. And then the silicon substrate was sealed with Pyrex glass by using a diffusive bonding technique. By means of the microfilters located inside microchannels, RBCs and other microparticles were effectively separated.

Then Wilding [16] furthered this method of microfiltration and developed another microfilter for separating WBCs from whole blood samples *via* silicon-glass hybrid microchips. The silicon substrate consists of a series of micropillars (Fig. **1A**, **B** and **D**) or "weir-type" filters (Fig. **1C**) as cellular trapping spots. The microfilters were fabricated inside a silicon microchamber by MEMS technologies. The filtered WBCs were trapped on the microfilters and used for the PCR reactions subsequently.

Manual counting of cells trapped on the filtration pillars in the weir-type filter chip (Fig. **1C**) revealed that the microchip isolated approximately 1200 WBCs. In human blood approximately 34% of WBCs are lymphocytes (6-15 μm diameter) and these are smaller than the polymorphonuclear cells (*e.g.*, neutrophils, average diameter 12 μm; eosinophils, average diameter 13 μm) and monocytes (14-20 μm). RBCs at rest assume a biconcave discoid shape with a diameter of ~8 μm and a thickness of ~2 μm passed through the microstructure barrier, while larger WBCs with a diameter of more than 10 μm were trapped by them. By using the weir-type filter chip (Fig. **1C**), WBC isolation yields ranged from 4 to 15%.

Moreover, micro structures including micropillar arrays, microweir structures, membranes with circular, hexagonal and rectangular *via* holes, which were micromachined by a composite silicon nitride/parylene membrane method based on MEMS technologies, were used as particle filters for the gaseous samples [21]. Cell/particle separation by using microstructure barriers is not only a simple and non-destructive method but can be integrated with other components of microfluidics.

At the same time, there are problems of clogging or jamming in most of these separation microchips due to dead-end filtration. For dead-end filtration, the fluidic flow is perpendicular to the filtration structures so that smaller particles pass through the filtration barriers along the fluid, while larger particles are stopped to cause channel clogging or jamming [23].

To overcome clogging of the dead-end filtration structure, crossflow filtration designs, which allow the bigger particles to stay in a suspended state instead of being trapped after being stopped, have been used in microfluidic chips easily.

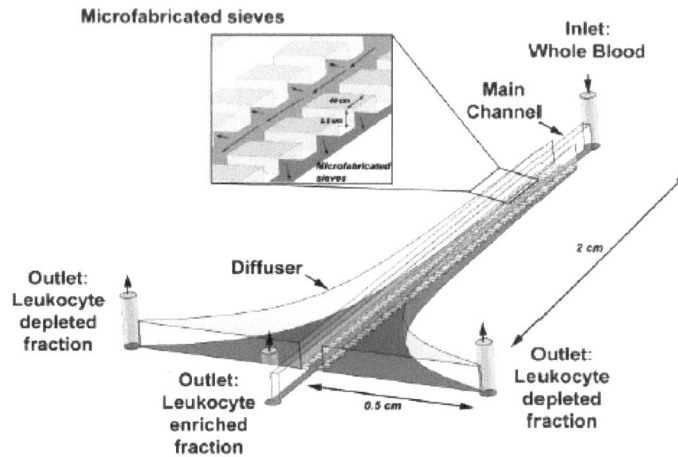

Figure 2: Schematic of the diffusive filter for size based continuous flow fractionation of erythrocytes from whole blood. Insert shows the 40μm × 2.5μm sieve structure and the arrangement connecting the main channel to the diffuser [22]. Reprinted with permission from the Royal Society of Chemistry. Copyright (2006).

Brody *et al.* [24] firstly suggested the separation of plasma from whole blood using a planar microfabricated filter device *via* tangential flows, and reported the filtration of a suspension of microspheres. Toner and coworkers [22] developed a crossflow filtration microdevice with a glass-PDMS hybrid microchip for leukapheresis. As one subgroup of Apheresis, leukapheresis is a procedure to selectively remove leukocytes for blood transfusions, which can avoid some adverse effects [25] like febrile transfusion reactions, graft versus host disease, transmission of infectious agents like viruses (cytomegalovirus CMV), herpes virus, T-cell leukaemia/lymphoma virus), bacteria, toxoplasma gondii, prions, platelet refractoriness and transfusion related immunomodulation. As shown in (Fig. **2**), the microchip was fabricated by simple soft lithographic techniques in which micro sieves were used to exploit the size and shape difference between the different cell types to obtain depletion of leukocytes from whole blood. For the given device design, isolation of ~50% of the inlet RBCs, along with depletion of >97% of the inlet WBCs was reported by optimizing flow rates.

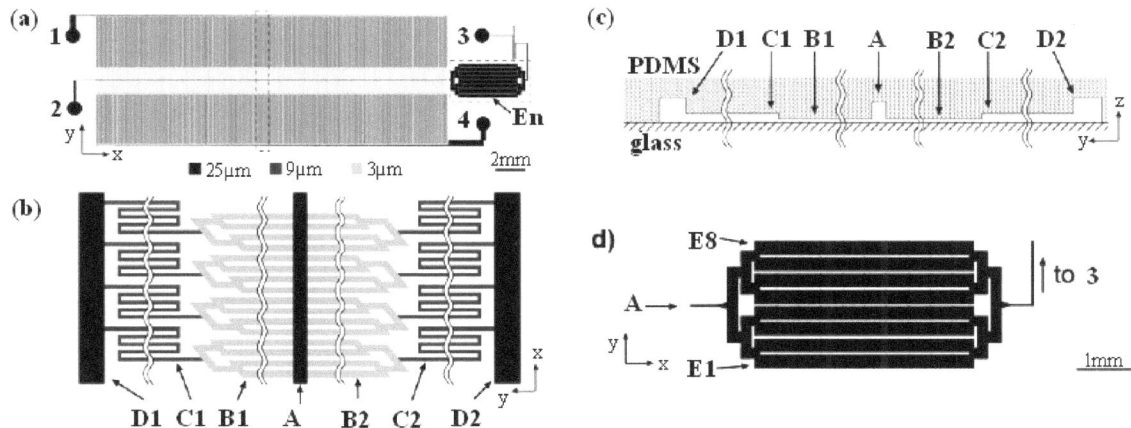

Figure 3: (a) Drawing of the microfluidic device. Ports labeled 1-4 are blood inlet, perfusion inlet, WBC outlet, and RBC outlet, respectively. **(b)** Blowup of a fragment of the separation network outlined with dotted line in (a) turned counterclockwise by 90° with respect to (a). **(c)** Cross-sectional view of channels in the separation network. Dimensions are not to scale. **(d)** Blowup of E channels outlined with dotted line in (a). Channel depths, 25, 9, and 3μm, are gray scale coded in (a), (b), and (d) [26]. Reprinted with permission from the American Chemical Society. Copyright (2007).

VanDelinder and Groisman [26] described a microfluidic device to separate WBCs from whole blood in a cross-flow manner, which consists of an array of microchannels with a deep main channel and large number of orthogonal, shallow side channels, shown in (Fig. **3**). As a suspension of particles advance through the main channel, a perfusion flow through the side channels gradually exchanges the medium of the suspension and washes away particles that are sufficiently small to enter the shallow side channels. This device was shown to reduce the content of RBCs by a factor of ~4000 with retention of 98% of WBCs. The proposed technique of separation by perfusion in continuous cross-flow could be used to enrich rare populations of cells based on differences in size, shape, and deformability.

Using an ion beam track etching technique, Metz [27] developed a polyimide microfluidic chip for selective delivery or probing of fluids to biological tissues. They created submicron pores on the selected areas of the microchannels by high-energy, heavy ions irradiation and chemical etching. However, the crossflow filtration experiments were only carried out with fluorescent polystyrene beads that differ significantly in size, and not implemented with real biological samples. Separation limits were mainly determined by the pore size distributions of the ion track membrane.

3.2. Optical-Based Separation

Fluorescence-Activated Cell Sorting (FACS) is one of the most common methods based on optical identification used to evaluate cell population. Despite its effective performance, it is not widely used in biological laboratories because of its high cost.

The group of Quake *et al.* [28] developed micro Fluorescence Activated Cell Sorters (μFACS) which can significantly reduce the cost of the macro-scale FACS. The μFACS device is a PDMS microfluidic chip with T-type channels fabricated by the soft lithography technique. The chip is mounted on an operated system including an inverted optical microscope, a light source of laser, a photomultiplier tube (PMT) and a computer, shown in (Fig. **4**). The signal of the potential fluorescence of particles excited by the laser is displayed on the screen of the computer. And the particle samples are dispensed, interrogated, sorted and recovered by using electrokinetic flow controlled by three platinum electrodes at the input and output wells.

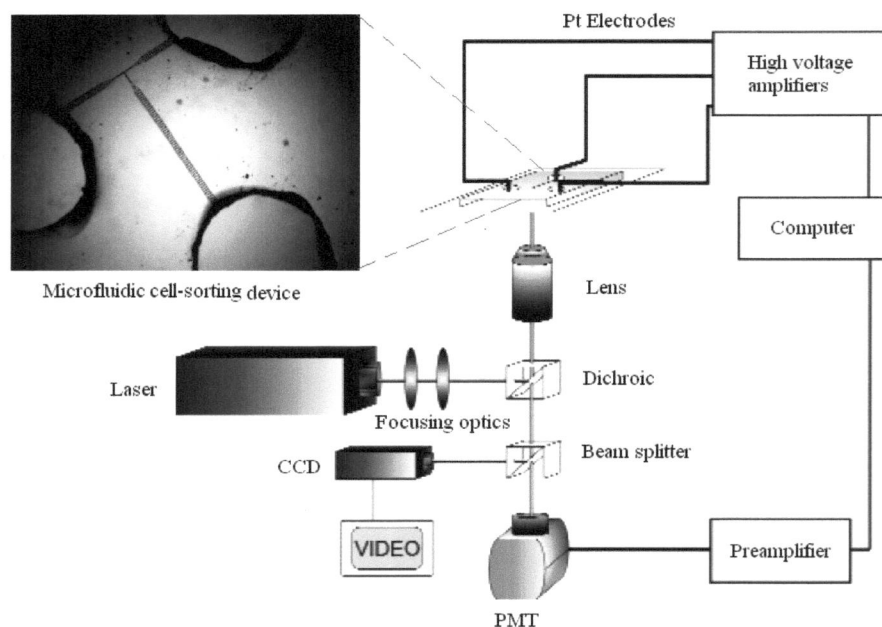

Figure 4: Schematic diagram of the cell sorting apparatus. Optical micrograph of the μFACS device (insert). The device shown has channels that are 100 μm wide at the wells, narrowing to 3 μm at the sorting junction. The channel depth is 4 μm, and the wells are 2 mm in diameter [28]. Reprinted with permission from the Nature Publishing Group. Copyright (1999).

Then the same group [29] furthered the idea of micro fluorescence activated cell sorters and developed an integrated microfabricated cell sorter with integrated pneumatically activated pumps and valves. The authors employed the multilayer soft lithography technology to create glass/PDMS hybrid microfabricated fluorescent activated cell sorting devices. The cells are driven and diverted into a collection chamber by using the micropump and microvalve system, after the special cell is identified on the basis of their fluorescent properties by using the optical system. For a given typical cell sorter chip, cells can be sorted at rates of up to ~40 cells/s with enrichment factors of ~90 and recovery yields between 16 and 50%.

Since the Quake group reported the first μFACS device [28] in 1999, there are a few types of controlling principles reported afterwards for μFACS such as hydrodynamic-driven [29-31], electroosmotic-driven [32, 33], electrokinetic-driven [34], and dielectrophoretic-driven [35] principles. The process of cell separation is described briefly as follows. First, the optical system identifies the type of the cell with different fluorescence. Then the signal of the determined cell type is sent to the driven system with one or two of above controlling principles. Finally, the focused cell pathway is deflected and collected at its predicted outlet.

Wolff and colleagues [36] reported a high-throughput μFACS integrated with various advanced functionalities, shown in (Fig. **5**). The functionalities include a novel microfluidic structure for sheathing and hydrodynamic focusing of the cell-sample stream, a chip-integrated chamber for holding and culturing the sorted cells, and integrated optics for detection of cells. As a part of the sorting chip, a monolithically integrated single step coaxial flow compound for hydrodynamic focusing of samples in flow cytometry and cell sorting was developed. The structure is simple, and can easily be microfabricated and integrated with other microfluidic components. An integrated chamber on the chip was designed for holding and culturing the sorted cells. By integrating this chamber, the risk of losing cells during cell handling processes was eliminated. The sheathing and hydrodynamic focusing of samples were used to sort cells by pressure-driven fluid flow. Using this sorter, fluorescent latex beads were sorted from chicken RBCs. A much higher sample throughput of 12000 cells/s with 100-fold enrichment than previous μFACS was reported. These designs of the μFACS provide an inexpensive, robust, and flexible way to sort and manipulate single cells.

Although the FACS generally requires fluorescence labeling, the μFACS that is without labels can be achieved in certain cases such as auto-fluorescence property of cells [37]. Two different populations of human blood cells (granulocytes and RBCs) were separated from each other based on their intrinsic fluorescence patterns by using a μFACS microfluidic chip and conventional Electroosmotic Flow (EOF) to switch the direction of the fluid flow. The μFACS microfluidic chip was just a simple microfluidic three-port glass microstructure. This method can minimize the potential risk of damaging cells due to the effects of fluorescent labels, and cuts down the sample preparation time and cost.

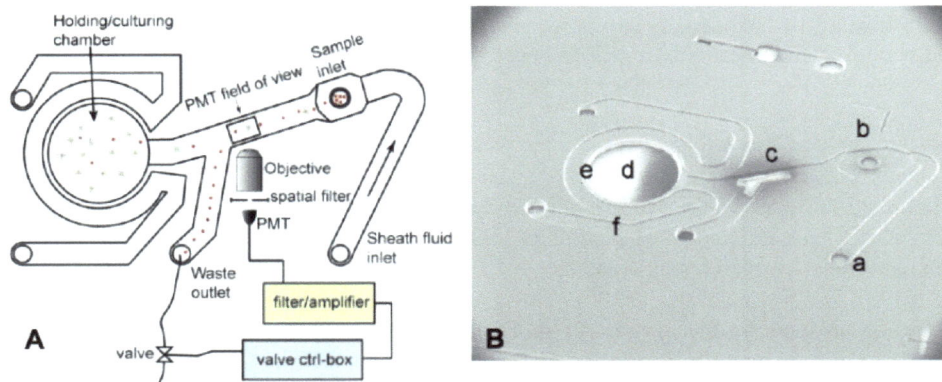

Figure 5: Schematic set-up for the second generation μFACS. B: Scanning Electron Microscope (SEM) image of the second-generation micro cell sorter chip with integrated holding/culturing chamber: **(a)** sheathing buffer inlet, **(b)** "chimney" sample inlet, **(c)** detection zone, **(d)** holding/culturing chamber, **(e)** sieve to allow diffusion of nutrients and confinement of cells, **(f)** channel for draining excess liquid during sorting and for feeding fresh media to the cells during cultivation, **(g)** waste outlet [36]. Reprinted with permission from the Royal Society of Chemistry. Copyright (2003).

Another non-invasive method for cell sorting is to use optical forces to modulate the direction of the flow. For example, Wang *et al.* [38] implemented a microfabricated cell sorter for mammalian cell separation using optical forces instead of mechanical valves to switch the direction of cells,. This µFACS reported a slightly higher throughput of ~100 cells/s with the recovery yield above 85% and the enrichment factor up to ~70.

The group of Grier demonstrated that periodic potential landscapes can be used to separate mesoscopic objects [39]. They proved the possibility of optical fraction which was widely used in the recent advances in manipulation of optical tweezers. Optical tweezers are manipulation tools developed by Ashkin, in which a tightly focused single laser beam is used to trap a single particle [40]. Optical tweezers have been widely used since then to trap single particles/cells in microfluidic devices.

MacDonald [41] used optical forces to sort microscopic particles in microdevices. Particles experiencing sufficient optical forces are kinetically locked in arrays of optical tweezers, whereas other particles flow along the natural stream, as shown in (Fig. **6**). In optical fractionation, an optical gradient force called potential energy landscape is generated to deflect particles or cells from their natural pathway according to their size or other intrinsic properties.

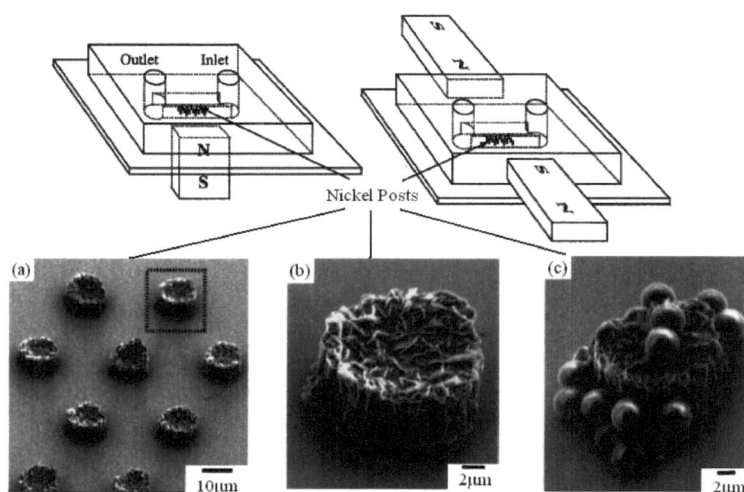

Figure 6: Schematic of microfiltration system after the integration of the nickel posts into a PDMS fluidic channel [46]. **(a)** SEM of a typical array of nickel posts. The post was ~7µm in height and ~15µm in diameter. The spacing between two neighboring posts was ~40µm and the spacing between two neighboring arrays was also ~40µm; **(b)** SEM of magnified view of a post. The edge roughness was ~1µm and the surface-roughness was ~500 nm; **(c)** SEM of a post attached with magnetic beads after drying and separation from the PDMS channel. The beads were ~4.5µm in diameter. Reprinted with permission from the American Institute of Physics. Copyright (2002).

Then the sorting method based on optical lattices was developed [42-44] with the separation efficiency up to 95%. The particle clogging or jamming was decreased by using flashing lattices during the sorting process. This separation method based on optical gradient forces has an advantage over conventional µFACS. The purity and efficiency of sorting is higher since particles/cells can be flushed through the landscape in parallel rather than a single channel.

Optical separation presents for the most part major advantages in terms of sensitivity, selectivity as well as versatility and permits the sorting of particles/cells in a continuous flow. But there are still challenges for miniaturizing the pneumatic control setup or optical system.

In addition, the need for laser sources hinders the easy portability of such systems although the integration of vertical cavity surface emitting lasers arrays offers real opportunities [45]. A real need for miniaturization of the apparatus is necessary in this field. The scale up of such optical and fluidics microsystems is another concern.

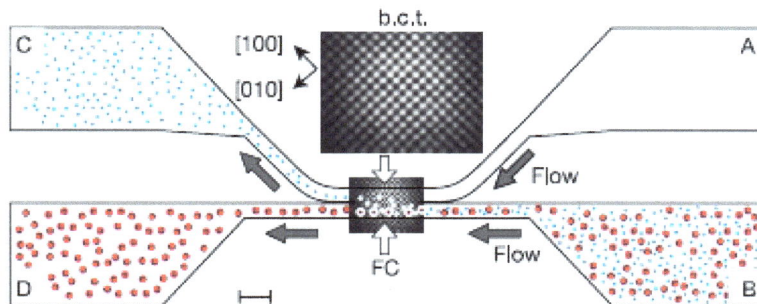

Figure 7: The concept of optical fractionation [41]. Low Reynolds number flows will be laminar: without an actuator all particles from chamber B would flow into chamber D. Chamber A would typically introduce a 'blank' flow stream, although this could be any stream into which the selected particles are to be introduced. By introducing a three-dimensional optical lattice—in this case a body-centred tetragonal (b.c.t.) lattice—into the Fractionation Chamber (FC), one species of particle is selectively pushed into the upper flow field. The reconfigurability of the optical lattice allows for dynamic updating of selection criteria. For weakly segregated species, the analyte can be either recirculated through the optical lattice or directed through cascaded separation chambers. This latter option also allows the use of multiple selection criteria in a single integrated chip. The flow volume in our current sample cells is 100 mm thick; scale bar, 40 mm. Reprinted with permission from the Nature Publishing Group. Copyright (2003).

3.3. Magnetic-Based Separation

Magnetic Activated Cell Separation (MACS) is one of the most useful tools in biotechnology to separate cells of interest out of mixed cell populations. In the magnetic sorting technique, separating particles/cells can be sorted into two types. One type is the non-immunological magnetic separation using inherent paramagnetic property of different cells. Another type is the immunological magnetic separation by using the labeled magnetic micro/nano beads.

The magnetic separation methods were initially developed on larger scales [13, 17, 47-49], where a magnet was placed in the vicinity of a column containing the cells to be separated. The separation process is as follows. First the cells of interest are labeled with magnetic micro/nano beads, after the magnetic beads are modified with various different specific antibody proteins that are specific to the cell membrane protein of interest. Then the mixture sample of both labeled cells and non-labeled cells are loaded into a column. When the magnet field is switched on, the magnetically labeled cells are retained in the column, whereas non-labeled cells are flushed away with the buffer. Finally, the sorted cells are collected by the flushing buffer after the magnet field is switched off. The efficiency rate of the protocol is often higher than 95% [50].

Magnetic particles are applied to sort cells based on magnetic force with many physical approaches such as magnetic susceptibility [18, 51, 52], binding capacity [53-55] and separation using cell samples [18, 56], which have been developed by several groups.

Although the method of MACS has been widely researched and applied commercially [57], it is still a batch process. It is difficult for the commercial MACS to separate and collect different cells continuously. To cut down the volume and cost of the large commercial apparatus and continuously collect the sorted cells, the micro MACS has been brought forward and developed by many research groups based on microfluidic techniques and MEMS technologies. In the method of miniaturized MACS, the magnetic force becomes stronger and more precise due to the fact that the magnets are closer to the microchannels where cells flow through. Moreover the μMACS requires only few quantities of samples and lower cost because of its small size compared with the conventional macro magnetic filtration systems.

In these earlier examples, small permanent magnets were fabricated in microfluidic devices and used for magnetic particle trapping. Deng *et al.* [46] developed a miniaturized magnetic sorting system integrated with arrays of micron-scale nickel posts, shown in (Fig. **7**). These nickel posts were fabricated into a PDMS microfluidic channel by using the soft lithography technique and the electro deposition technique. The posts are used as magnetic elements to trap magnetic particles. The separation process is just similar with its macro equivalent in which non-magnetic beads and magnetic beads are sorted and collected into "Solution 1" and "Solution 2" by flowing buffers in turn. All nonmagnetic beads were collected in "Solution 1" while more than 95% of the magnetic beads were collected in "Solution 2" and less than 5% collected in "Solution 1". Microfabricated magnets in close proximity to the channel can generate high magnetic field gradients, but have a limited reach and might lead to extensive heating.

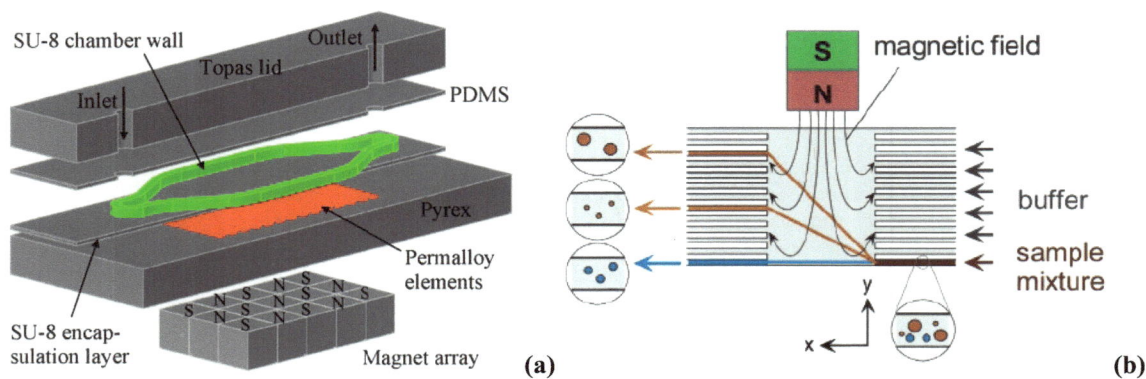

Figure 8: (a) Exploded view of the magnetic bead separator[58]. Reprinted with permission from the Elsevier. Copyright (2008); **(b)** Concept of free-flow magnetophoresis [59]. Magnetic particles are pumped into a laminar flow chamber; a magnetic field is applied perpendicular to the direction of flow. Particles deviate from the direction of laminar flow according to their size and magnetic susceptibility and are thus separated from each other and from nonmagnetic material. Reprinted with permission from the American Chemical Society. Copyright (2004).

It is possible to separate RBCs from whole blood by the magnetic method without magnetic particles because of hemoglobin. Hemoglobin is a conjugated metal-protein comprising of four polypeptide globins chains containing a ring structure covalently bonded with a central ferrous iron atom (Fe^{2+}), which binds reversibly with oxygen. When deoxygenated, each of the four iron atoms contains four unpaired electrons, giving the protein and the cell a substantial paramagnetic moment. WBCs that do not contain hemoglobin are diamagnetic particles. These properties are the starting point for the trapping of RBCs particles using high gradient magnetic separation. By leveraging this unique property of RBCs, Iliescu [60-62] reported a glass microfluidic device with ferromagnetic "dots" for sorting RBCs from blood under a continuous flow.

In Iliescu's research, the device consists of a glass substrate with ferromagnetic "dots" and a glass cover with the inlet/outlet holes and a microchannel. The cover was fabricated by twice wet etching in HF/HCl. The substrate was fabricated by a simple lift off process and the metal "dots" Ti/Ni (50 nm/2 μm) were deposited on its bottom. The geometry of the ferromagnetic "dots" and the application of an external magnetic field perpendicular to the flow direction generate a magnetic force on the RBCs present in the blood. During the diluted blood passing through the microfluidic channel, an average of 95 % of RBCs were captured on the bottom of the microfluidic channel under the gradient of magnetic field, while the rest of the blood was flushed out and collected at the outlet.

Bu *et al.* [58] presented an efficient magnetic bead separator based on an external checkerboard array of permanent magnets providing long-range magnetic capturing forces combined with on-chip permalloy elements providing strong short-range magnetic retaining forces, shown in Fig. 8(a). Although this device wasn't applied to real samples, it has provided a technique for a high-throughput application, *e.g.*, the separation and trapping of rare biological samples from a large volume of raw samples. Furthermore, the Reich group [63-65] reported a method of non-immunological magnetic separation by using magnetic nanowire to bind to cells, and therefore give cell magnetic properties for cell separation.

Some attempts have also been tried to fabricate continuous flow magnetic separation microfluidic devices. For example, a technique called 'on-chip free-flow magnetophoresis was demonstrated in literatures [59, 66-69]. Xia [70] reported an integrated, microfluidic device with a microfabricated NiFe comb and an external magnet to separate magnetic particles from non-magnetic particles by using a local magnetic field gradient. The cells bound to the magnetic particles lie in an inhomogeneous magnetic field perpendicular to the direction of flow and then the labeled cells are changed their pathway in magnetic field, while the cells not bound to the magnetic particles have no influence on the magnetic field and keep their pathways. Thus the labeled cells and non-labeled cells can be separated during continuous flowing fluids without any wash steps.

In this design, a high gradient magnetic field concentrator was microfabricated at one side of a microfluidic channel with two inlets and outlets. The hybrid magnets of a larger external magnet and a small magnetisable feature in close proximity to the microchannel can obtain a higher gradient magnetic field, resulting in a higher flow rate for the cell separation. The most efficient magnet setup for continuous flow separations is a combination of magnetized NiFe microcomb and an external permanent neodymium magnet. Under the optimal condition, the separation efficiency of *E. coli* from the fluids containing RBCs was about 78%. The throughput of the micromagnetic separator was about 10,000 cell/s for the sample with a high cell concentration.

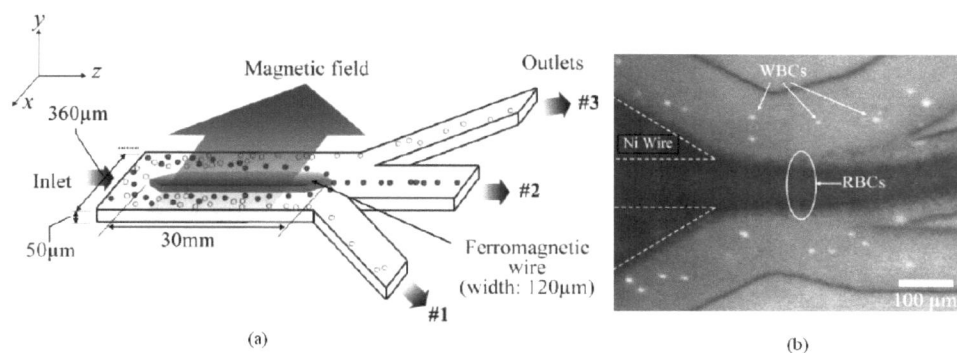

Figure 9: (a) Illustrations of the single-stage PMC magnetophoretic microseparator with a rectangular ferromagnetic wire. Perspective view of the microchannel that has one inlet and three outlets. **(b)** Fluorescently probed WBCs passing through the microchannel of the three-stage cascade PMC microseparator at an average flow velocity of 0.1 mm/s [71-73]. Reprinted with permission from the Royal Society of Chemistry. Copyright (2006).

Pamme and Manz [59] developed a microfluidic chip which could be beneficial not only to separate magnetic particles from nonmagnetic material but also to separate magnetic particles with different properties from each other. The glass microchip features a separation chamber with a number of inlet and outlet channels (Fig. **8b**), which was fabricated by a direct write laser lithography system. A micromagnet placed upon the channel provides a non-homogeneous magnetic field which is perpendicular to the direction of flow. The mixture of different magnetic particles and non-magnetic particles is aligned along the wall of the microchannel. Depending on their size and magnetic properties, particles are deflected more or less from their original pathways. The addition of spacers further allows the collection of particles in separated outlets as shown in (Fig. **8b**).

A continuous flow, magnetic separation device for the enrichment of fetal cells from maternal blood was described in a patent by Blankenstein [74]. This device uses the same working principle as the one reported by Pamme although the cells are required to be labeled. Another type of continuous magnetic separation was demonstrated by Inglis *et al.* [75, 76]. In these examples, some ferromagnetic strips fabricated in microchannels provide an array-like magnetic field pattern at a given angle to flow direction. Cells selectively tagged with magnetic nanoparticles deflect from the flow path to follow the microfabricated magnetic strips. This technique convincingly demonstrates the separation of leukocytes from human blood. A sequence of images were presented to illustrate the difference in the path of a tagged leukocyte from that of untagged RBCs. Continuous cell by cell separation from a flow stream by selectively tagging with magnetic beads was realized by using a lateral magnetic force on streaming tagged cells.

In another example, nickel wires in close proximity to the microchannel were magnetized by an external permanent magnet for the separation of RBCs in a continuous flow manner reported by Han [71-73]. An external magnetic field was used to activate a ferromagnetic wire integrated to a microchannel as shown in (Fig. **9**). By using the high gradient magnetic separation method, continuous single-stage and three-stage cascade paramagnetic capture magnetophoretic microseparators were designed and successfully demonstrated for directly separating RBCs and WBCs from whole blood based on their native magnetic properties. 91.1% of RBCs were continuously separated from the sample by using the single-stage paramagnetic capture microseparator. 93.5% of RBCs and 97.4% of WBCs were continuously sorted from whole blood by using the three-stage cascade paramagnetic capture microseparator.

Jung and Han [77] reported another glass microchip by using the same working principle of the lateral-driven continuous magnetophoretic separation of red and white blood cells from peripheral whole blood, based on their native magnetic properties. A ferromagnetic wire array was laid on glass substrate. The wire array created an even lateral magnetophoretic force on the whole area of the microchannel, improving the separation efficiency and throughput. The designs proposed by Han and his colleagues belong to non-immunological magnetic separation using inherent paramagnetic property of cells. This approach enables separation of blood cells without the use of additives such as magnetic beads. Different cells were repelled or attracted by the wire and thus collected at different outlets, depending on their internal properties.

3.4. Affinity-Based Separation

Conventional cell separation can be realized by immunoreactions of membrane proteins with the capturing antibodies for specificity. The immunological technique is a mainstay of commercialized cell separation methods such as the fluorescence-activated cell sorting and magnetic-activated cell sorting.

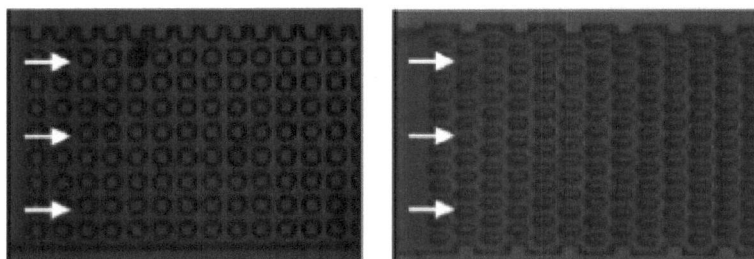

Figure 10: Microstructured flow channels (top views) [78]. Square (left) and Offset (right). The structures were fabricated with Deep Reactive Ion Etching (DRIE). These views of the microchannels are seen through the top coverglass. Downstream flow is in the left-to-right direction, and "height" is along the axis into the image. The depth of the channels was 40 mm. Reprinted with permission from the Royal Society of Chemistry. Copyright (2005).

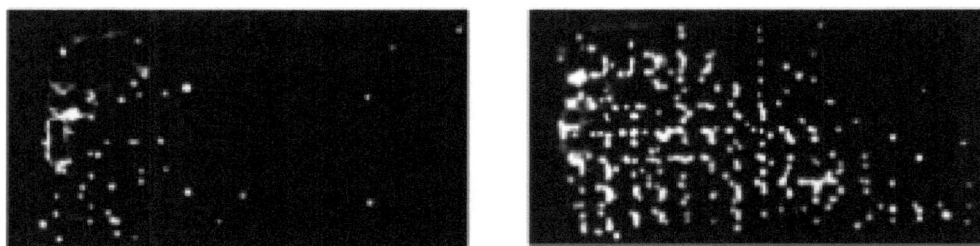

Figure 11: HL-60 cells captured on the Square lattice [78]. Two views taken 2 min apart show cells (white) accumulating on the surfaces of the microchannel. Cells were captured from a flowing suspension and transited slowly down the length of the channel (flow from left to right). Reprinted with permission from the Royal Society of Chemistry. Copyright (2005).

Affinity-based cell separation uses solid supports where the specific antibodies are conjugated. If the heterogeneous cell suspension filters through the antibody-bound solid phase, cells of interest are captured on the solid supports while other contaminants would pass through the way. The bound cells can be released by changing the buffer solution.

The biggest advantage of this approach is high specificity and selectivity due to the nature of immunoreaction between the membrane marker proteins and labeling antibodies. But it also has several disadvantages such as the potential damage of immunologically isolated cells [79], complicated processes and high cost, since expensive antibodies are involved.

In order to avoid the disadvantages of the conventional affinity-based separation, some microfabricated devices for affinity-based cell separation have been introduced with some unique functions compared to their macro counterparts. The microfabricated tools can reduce the cost and separation time compared with the conventional affinity cell separation technique.

For example, cells can be separated on the basis of the affinity of cell-surface receptors for proteins immobilized on the surfaces of the microfluidic channel. Revzin and co-workers described a microfabricated cytometry platform for cell sorting and characterization, providing a high-density leukocyte matrix array for isolating, characterizing and releasing the cells [80]. Poly (ethylene glycol) (PEG) photolithography was used to fabricate arrays of microwells composed of PEG hydrogel walls and glass attachment pads of 20 mm×20 mm and 15 mm×15 mm in size. PEG micropatterned glass surfaces were further modified with cell-adhesive ligands, poly-L-lysine, anti-CD5 and anti-CD19 antibodies. The specific cell-surface interactions between the specific group of antibodies chemically fixed on the glass surface and the cell of interest was implemented within the individual wells. Glass slides micropatterned with PEG and cell adhesive ligands were exposed to T-lymphocytes for 30 min. These anchorage-independent cells became selectively captured in the ligand-modified microwells forming high-density cell arrays. Cell occupancy in the microwells was found to be antibody-dependent, reaching 94.6±2.3% for microwells decorated with T-cell specific anti-CD5 antibodies.

Similarly, Chang demonstrated adhesion-based collection and separation in a microfluidic channel by mimicking the physiological behavior of leukocyte in blood vessel [78]. The "separation columns" consists of microfluidic flow channels containing micropillars to maximize surface area for capture of passing cells, shown in (Fig. **10**). The provision for micropillar arrays maximizes the surface area presented to the passing cells. Two different microstructures were studied: a ''square array'' of 25 × 25 μm square posts, spaced 25 μm apart; and an alternating array, called the ''offset array'', of thinner pillars, spaced 30 μm but with successive downstream rows offset 15 μm in the cross-stream direction, shown in (Fig. **10**). Test structures were fabricated in silicon using deep Reactive Ion Etching (DRIE).

A water-soluble human E-selectin IgG chimera was used as the adhesion protein for higher specific capturing. An unusual property of selectin-mediated adhesion is the possibility of sudden cell arrest from rapid transit in the fluid suspension due to high bond association rates, which allow selectin and ligand to bond after very brief though close apposition. Capture and arrest of HL-60 cells occurred along surfaces in both of the microchannel designs, shown in (Fig. **11**). Transient adhesion between cells and appropriate surface ligands retard cell movement through the channel. HL-60 cells were separated from U937 cells on the basis of differences in their retardation, albeit with low resolution.

The group of Toner [81] reported a microfluidic device that can efficiently and reproducibly isolate Circulating Tumor Cells (CTCs) from the blood of patients with common epithelial tumors. The CTC-chip consists of an array of microposts chemically modified with anti-epithelial-cell-adhesion-molecule antibodies which provides the specificity for CTC capture from unfractionated blood. The CTC-chip successfully identified CTCs in the peripheral blood of patients with metastatic lung, prostate, pancreatic, breast and colon cancer in 115 of 116 (99%) samples, with a range of 5~1,281 CTCs per ml and approximately 50% purity. In addition, this CTC-chip is able to sort rare cells directly from whole blood in a single step by using modified microposts, comparing to magnetic-bead-based approaches.

A similar microfluidic device was developed to identify and specifically collect tumor cells of low abundance (1 tumor cell among 10^7 normal blood cells) from circulating whole blood [82]. By immobilizing anti-EpCAM (Epithelial Cell Adhesion Molecule) antibodies on polymer micro-channel walls by the chemical surface modification of PMMA, breast cancer cells from the cell line MCF-7, which over-express EpCAM on their surfaces, were caught by the strong binding affinity between the antibody and antigen.

In this design, three EpCAM/Anti-EpCAM binding models were used to determine an optimal flow velocity, 2 mm/sec, which was capable of binding the maximum number of cells with an optimized critical binding force, and a maximum throughput. At higher velocities, shear forces (> 0.48 dyne) broke existing bonds and prevented formation of new ones. This detection micro device can be assembled with other lab-on-a-chip components for follow-up gene and protein marker analysis.

Additionally, the separation of sperm and epithelial cells in microfabricated devices was reported by Horsman using the epithelial cell property of adherence to the glass substrate [83]. The sperm cells were separated from the epithelial cell-containing biological mixture by using lower flow rates.

3.5. Ultrasonic Separation

High frequency, acoustic standing waves can be used to separate materials with different acoustic impedances based on their size, density and compressibility where particles tend to move to the pressure node of the standing wave. If the particles/cells are migrating in the acoustic field, the migration velocity is determined by the inner characteristics of the particles/cells. The method is shown to be suitable for both biological and non-biological suspended particles. Particle separation is accomplished by combining laminar flow with the axial acoustic primary radiation force in an ultrasonic standing wave field. The conventional ultrasound systems for cell or particle separation have been variously studied before the emergence of microfluidic ultrasonic devices.

Recently, continuous flow separation of particles has been performed with microfluidic ultrasonic devices. Coakley [84-87] and Nillsson [88-90] reported several papers for suspended particle separation in microfabricated chips. The acoustic force on a particle depends on the frequency and amplitude of the acoustic field, besides the properties of the particle and its surrounding medium. The higher the frequency and the larger the applied voltage, the higher the acoustic force is. The required frequency is predetermined by the width of the microchannel. For example, a channel of 1 mm width requires a wave frequency of about 100 kHz, whereas a channel of 10 mm width requires 10 MHz. In short, the smaller the channel, the higher the required frequency and in turn the larger the acoustic force generated. For microfluidic ultrasonic devices, a standing ultrasonic sound wave is usually generated by micro piezoelectric transducers and be over the cross section of a microchannel. Particles or cells subjected to the sound wave experience a force, either towards the node or towards the anti-node.

Cousins used the ultrasonic standing wave to separate plasma from whole blood in a tube [91]. Although plasma isolation was not implemented in microfluidic devices, it proved that the ultrasonic technology could be used for on-chip blood separation.

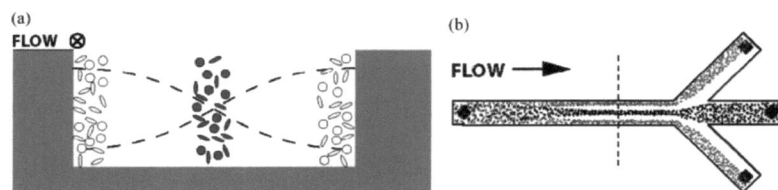

Figure 12: (a) Two particle types positioned, by the acoustic forces, in the pressure nodal and anti-nodal planes of a standing wave. (Cross section of the channel in (b), dashed line.) **(b)** Top view of a continuous separation of two particle types from each other and/or a fraction of their medium [92]. Reprinted with permission from the Royal Society of Chemistry. Copyright (2004).

Separation of lipid from blood in a microfluidic channel was described by Laurell's group [92]. Pressure fluctuations in a liquid medium result in acoustic radiation forces on suspended particles. Separation of particles is realized depending on different physical properties. One type of particle ends up in the pressure node and the other type in the pressure anti-nodes, shown in (Fig. **12a**). This effect can be used to separate different types of particles in a continuous laminar flow. Using a half wavelength acoustic standing wave, one type of particle is located in the middle of the channel and the other type along the side walls, shown in (Fig. **12b**). When the channel is split into the three outlets, one particle type exits through the centre outlet and the others through the side outlets.

Erythrocytes and lipid particles (triglycerides) in blood plasma were used to test the ultrasonic separation technology (Fig. **12**). Since erythrocytes and lipid particles have different compressibility and density, they move to different directions. In the experiments, erythrocytes and lipid suspended in blood plasma were introduced in the microchannel. The erythrocytes moved towards the centre of the channel and the lipid particles moved towards the side walls under a half wavelength standing wave field orthogonal to the direction of flow. The end of the channel is split into three outlet channels conducting the erythrocytes to the centre outlet and the lipid particles to the side outlets due to the laminar flow profile. The tritium labeled lipid particles being added to bovine blood were used to test this microfluidic device. More than 80% of the lipid particles were removed while approximately 70% of the erythrocytes were collected in one third of the original fluid volume.

Jönsson suggested the possible clinical implication of the lipid removing microfluidic device [93]. The microfluidic separation chip with a trifurcation and eight parallel channels was fabricated by using a standard silicon processing technique. The eight parallel channels are a prerequisite for increased throughput. The trifurcation was used to divide the flow after separation which was similar with the structure reported by Petersson [92].

The ultrasonic separation microfluidic chip was tested on real human shed mediastinal blood collected by cardiotomy suction during cardiac surgery [93]. The efficiency in terms of RBC recovery was measured and the safety of the method in terms of electrolyte changes was also investigated. Erythrocyte recovery, in terms of a separation ratio, varied between 68% and 91%. Minor electrolyte changes took place, where levels of sodium increased and levels of potassium and calcium decreased. This research proved thereby, that the ultrasonic separation technology can be used on human shed mediastinal blood with good separation efficiency.

Then the same group of Laurell [92, 94, 95] also used the ultrasonic standing wave forces to separate particles having different physical properties within the perfused microfluidic channel (*i.e.*, lipid vesicles were continuously separated from erythrocytes). Separation efficiency of polyamide spheres was up to 100% using the further miniaturization of the separation device under the optimal conditions. Then the same group reported a free flow acoustophoresis which is capable of continuous separation of mixed particle suspensions into multiple outlet fractions, shown in (Fig. **13a**). A mixture of different particles with different sizes was laterally translated to different regions of the laminar flow profile, which was split into multiple outlets for continuous fraction collection.

As shown in (Fig. **13b**), the microfluidic chip has two inlets, a main channel (370 μm wide, 125 μm deep, 30 mm long), eleven sub-channels and five outlets. One inlet is the particle-free medium inlet, located at the beginning of the separation channel, and the other is the sample inlet. The main channel is split into eleven sub-channels connected to five outlets. This configuration enables the extraction of up to six particle fractions.

Three sizes of polystyrene spheres (3 μm, 7 μm, and 10 μm) were introduced to the microfluidic chip with three active outlets and flowed through the rectangular cross-section separation channel. After a continuous separation experiment, 93% of the 10 μm particles were collected at the center outlet while 76% of the 7 μm and 87% of the 3 μm particles were collected at the second and third outlets, respectively. Using four outlets, a mixture of 2 μm, 5 μm, 8 μm, and 10 μm suspended polystyrene particles was separated with between 62 and 94% of each particle size ending up in separate fractions.

RBCs (~7 μm in diameter, ~2 μm thick, density ~1.100 g/mL) and platelets (2-4 μm in diameter, density ~1.058 g/mL) were separated in this microfluidic chip by using free flow acoustophoresis and medium-density manipulation techniques. While suspended in the saline solution (0.9 mg/mL) with nutrient additives (CsCl, 0.22 g/mL), the two blood components were affected similarly by the acoustic forces and were thus difficult to separate. 92% of the RBCs were collected at the first outlet while 99% of the platelets were collected at the second outlet.

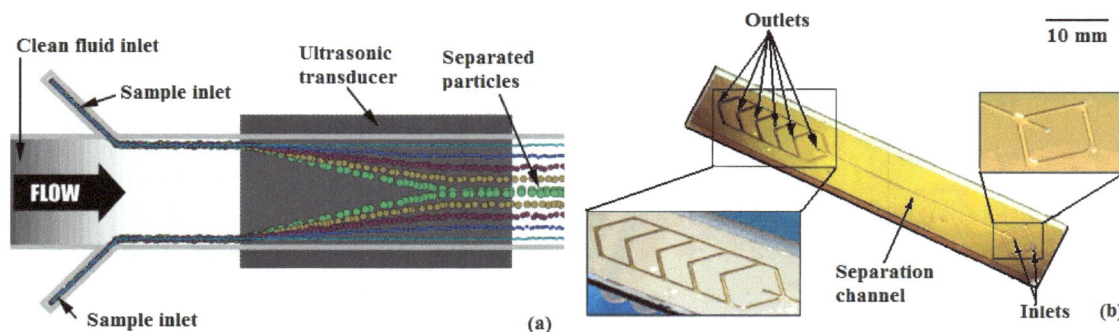

Figure 13: (a) Illustration of a particle suspension passing over the transducer where the particles are moved toward the center of the separation channel at a rate determined by their acoustic properties. Because of the laminar flow almost no mixing takes place. **(b)** Photograph showing the basic chip design that consists of a silicon wafer with wet-etched channel structures. The channels have been sealed with an anodically bonded glass sheet, and the inlets and outlets have silicon tubing connections on the back side of the chip [94]. Reprinted with permission from the American Chemical Society. Copyright (2007).

Whole blood containing RBCs, platelets, and WBCs (5-20 μm in diameter, density 1.062-1.082 g/mL) were separated by using this microfluidic chip with four active outlets. The saline solution with CsCl (0.22 g/mL) was used in the separation experiments. 79% of the platelets were collected at the fourth outlet; 86% of the RBCs were collected at the second and third outlets, while 92% of WBCs were collected at the third and fourth outlets. Although the separation results were not good enough for real applications, it proved that the medium manipulation, in combination with the free flow acoustophoresis, could be further used to enable the fractionation of RBCs, platelets, and WBCs.

Additionally, Petersson [96] reported suspended particle separation in a microfabricated chip and continuous blood medium exchanging device using ultrasound standing wave. The silicon substrate with a main rectangular cross-section channel (25 μm deep, 350 μm wide and 30 mm long) was prepared by using conventional anisotropic wet etching. At the beginning of the main channel, there are one center sub-channel and two side sub-channels originated from a common inlet. At the end of the main channel there are one center sub-channel and two side sub-channels with a common outlet. The silicon substrate with channels is sealed by anodic bonding of a glass cover with four holes. The separation microfluidic chip was ready for experiments, after four tubes and a piezoceramic plate were attached to the glass-side of the chip.

The sample mixture with particles enters through the side sub-channels while the clean solution enters through the center sub-channel, shown in (Fig. **14**). During the passage through the channel, the particles were observed to move from the contaminated sample solution into the clean medium in the middle of the channel and exit through the center outlet. This method can be furthered to exchange the carrier solution of RBCs. The washed RBCs can then be returned to the patient without risk of infusing potentially harmful substances.

As the first step, 5 μm polyamide spheres suspended in distilled water, spiked (contaminated) with Evans blue, were used to exchange their carrier medium for testing the microfluidic chip. More than 95% of the polyamide spheres were collected in the clean medium while removing up to 95% of the contaminant.

Then this microfluidic chip was used to wash blood. RBCs were switched from blood, spiked with Evans blue, to clean blood plasma. From separation experiments, 95% of the RBCs (bovine blood) were collected in clean blood plasma while up to 98% of the contaminant was removed.

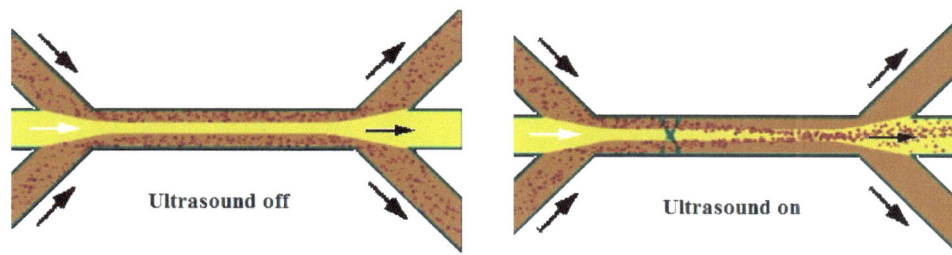

Figure 14: Schematic illustration of the medium exchange principle [96]. The particles in the contaminated medium enter through the side inlets and exit through the side outlets when the ultrasound is turned off (left). If the ultrasound is turned on, as indicated by the schematic standing wave in the right microfluidic structure, the particles are switched over to the clean medium and exit through the center outlet together with the clean medium, whereas the contaminated medium still flows to the side outlets. Reprinted with permission from the American Chemical Society. Copyright (2005).

Suspended particle separation using ultrasonic standing wave force has been widely used in traditional chemical engineering and material science. The elegance of acoustic separations stems from the fact that no physical contact is required between the ultrasonic transducer and the liquid. According to the literature, the ultrasonic treatment has not been shown to damage cells or biological material. The method, however, is not applicable to small nanoparticles or small molecules, given that the acoustic force is proportional to the cube of the particle radius. Furthermore, concerning the reported results on ultrasound cell separation especially in microfluidic devices, they have been usually fundamental and limited to separation of cells from obviously different types of materials such as lipids. Based on this current working principle, it seems that it is difficult to fractionate the cells of interest having subtle differences of size and surface marker expression.

3.6. Dielectrophoresis Separation

Dielectrophoresis (DEP) is a valuable method for manipulation of dielectric particles including polymer spheres, cells, proteins, and even DNA. Dielectric particles can be polarized along the direction of the electric field and the electrostatic forces on the two ends of the dipole are not equal, when they are put in an inhomogeneous electric field. Finally, the dielectrophoretic forces are produced because of the unequal electrostatic forces, resulting in the movements of the dielectric particles.

The dielectrophoretic forces depend on the difference in polarisability between particles/cells and surrounding medium. The forces also depend on the electrical properties of the particles, such as volume, as well as to the gradient of the electric field. For cells, the DEP forces are based on intrinsic electrical properties of cells, such as the internal structure, the size of the nucleus, or the bilipid membrane capacitance and conductance, both of which change with cell type and even with cell activation.

The DEP microfluidic devices have been widely studied due to the fact that electrical fields scale down favorably in chips and low voltages are enough to produce intense electrical fields. Moreover the thermal and hydrolysis effects which are harmful to living cells can be strongly reduced because of micro scale.

The DEP particle/cell separators have been developed using positive and negative DEP phenomena. When the particle exhibits a larger polarisability than the surrounding medium, positive DEP (p-DEP) occurs and p-DEP forces move the particles towards the stronger electric field, resulting in adsorption particles on electrodes. However, p-DEP forces for cell separation may cause reduced cell recovery [97, 98] and therefore they cannot be used for rare cell isolation and sorting.

When the particle is less polarisable than the surrounding buffer medium, negative DEP (n-DEP) occurs and n-DEP forces put the particle away from areas of high field intensity towards areas of low intensity. The strategy of n-DEP provides the repulsive forces acting on the cells, eliminating cell adsorption on the electrodes [99]. Although the n-DEP force is decreased as the cells are moving away from the electrodes, which gives rise to gradually diminished cell deflection velocity, the n-DEP is more adaptable to the continuous-flow cell separation because of its reliable cell recovery ratio.

Research on the DEP cell separation was first carried out using polymer beads as the analogue of cells in order to characterize the devices and principles [99-103]. And then the report from Yang demonstrated the possibility for cell separation [104].

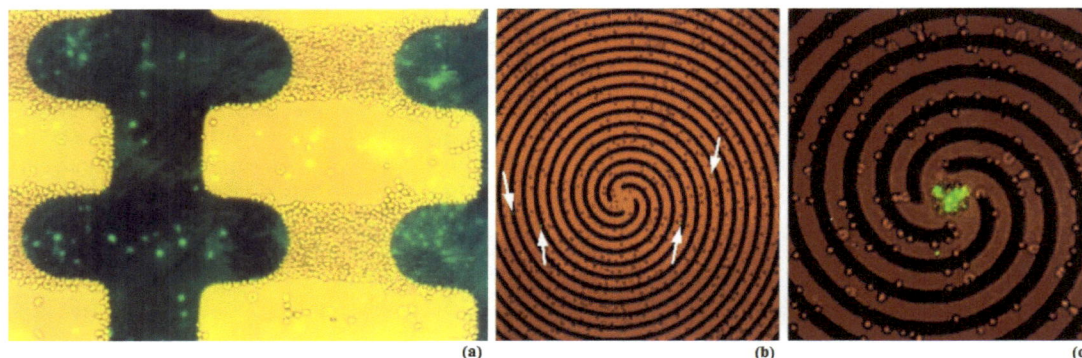

(a)   (b)   (c)

Figure 15: (a) Normal erythrocytes were trapped at the interdigitated electrode edges where there is the high electrical field, while 95% of parasitized cells exhibited a green fluorescence were repelled from the high field regions. **(b)** Prior to the application of an electrical field, parasitised cells (arrows) were spread throughout the sample. **(c)** Application of four phase signals to the spiral electrode elements caused normal erythrocytes to be trapped at the electrode edges while parasitised cells were levitated and carried towards the centre of the spiral by the n-DEP force [105]. Reprinted with permission from the Royal Society of Chemistry. Copyright (2002).

Gascoyne group reported various microfabricated DEP devices [105-108]. An interdigitated electrode array and the spiral electrodes were designed to exploit the dielectric differences between infected and uninfected cells and then to isolate malaria-infected cells from blood [105]. As shown in (Fig. **15 a**), the normal erythrocytes were trapped by positive dielectrophoresis in the inhomogeneous, high field regions between facing electrode tips while the parasitised cells were repelled into the gaps. Under these conditions, more than 99.5% of normal erythrocytes were trapped while 90% of parasitised cells remained free to be washed from the chamber by fluid flow. As shown in (Fig. **15 b** and **c**), normal erythrocytes were strongly trapped at the electrode edges while the parasitised cells were moved towards the centre of the spiral because of n-DEP. Approximately 90% of the parasitised cells and < 0.1% of the normal cells were focused to the centre. The parasitized cells were washed free by flowing suspending medium. Thus the infected and uninfected cells were separated from each other using DEP.

Huang described a microfabricated electrode array for isolating six types of cells by modulating DEP frequencies [109]. A 5 × 5 microelectronic array and a 10 × 10 microelectronic array were designed to separate five different cultivated cell lines and human peripheral blood mononuclear cells by using different DEP frequencies. Monocytic cells (U937), human T cell leukemia virus type 1 (HTLV-1), tax-transformed cells (Ind-2) from PBMC, as well as neuroblastoma cells (SH-SY5Y) from glioma cells (HTB) were separated. The purity of dielectrophoretically separated cells was reported higher than 95%.

Furthermore, cancer cells were isolated from normal blood cells by selective capture onto the dielectric affinity column containing a microelectrode array reported by Becker *et al.* [110]. The array of interleaved gold electrodes in the DEP device was designed to separate cancer cells from RBCs. Cell separation procedures were as follows. Firstly the mixture sample of the cancer cells and RBCs was introduced in the chamber. Then the elute flow was started at 5 μl/min and electrical signals were adjusted to capture cancer cells and then to release them. During the capture step, the lowering of the field frequency from 200 to 80 kHz was used to maximally capture tumor cells on the electrodes. At the same time, blood cells were released, eluted, and collected in the chamber outlet by using hydrodynamic forces. All tumor cells remained on the electrode tips under the p-DEP force and none was detected in stained slides of the eluted fraction. Finally the tumor cells were released by repetitively sweeping the applied field (twice per second) from 80 to 20 kHz. During the release step, the tumor cells were maximally purified.

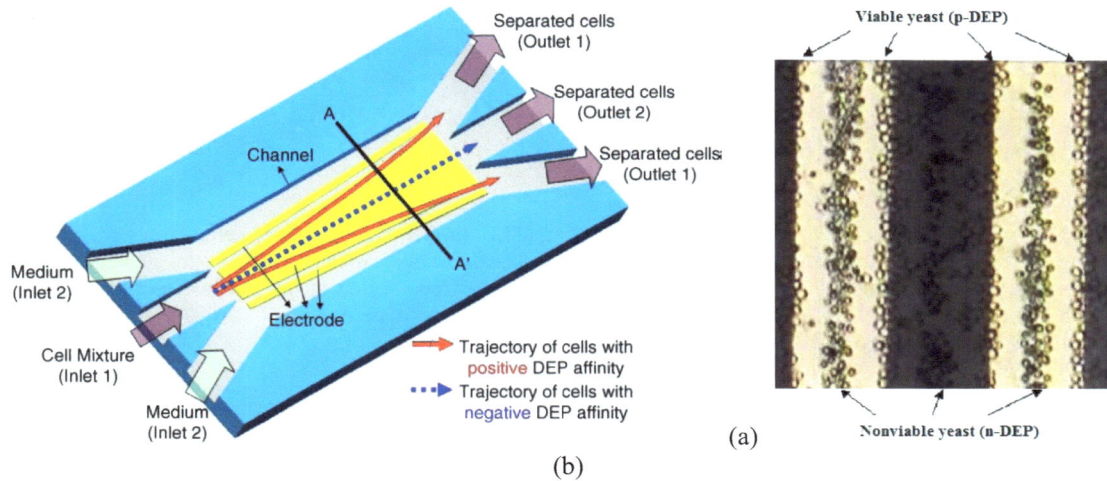

Figure 16: (a) Conceptual view of hydrodynamic dielectrophoresis (DEP) process. **(b)** DEP response measurement for the viable and nonviable yeast cells at the electric field frequency of 5 MHz [111]. Reprinted with permission from Elsevier. Copyright (2005).

Twelve minutes after application of the swept signal, the cell distribution on the electrodes was examined. With increasing distance from the inlet, the ratio of erythrocytes to MDA231 cells increased from 1:15 to 1:1 and remained at that value where the chamber was initially loaded with cells. Thereafter, the proportion of erythrocytes rapidly increased, reaching in excess of 99% at the outlet. After 20 min, retained tumor cell fractions were >95% pure.

Live and dead cells were also separated by interdigitated array electrodes [112] in continuous flow system [111]. In the continuous flow microfluidic device, live and dead yeast cells were continuously separated by appropriately balancing the hydrodynamic and dielectrophoretic forces acting on the cells reported by Doh [111], shown in (Fig. **16a**). Three planar electrodes were designed and fabricated in a separation channel. The positive DEP cells were flowed along with the streamline, while the negative DEP cells remained. In the experiment, the mixture of viable (live) and nonviable (dead) yeast cells was introduced in the mcirofluidic chip and was used to measure their DEP response by applying different electric fields (10 kHz and 5 MHz) and different medium conductivities (5, 30 and 78 μS/cm). Under the optimal conditions (the medium conductivity of 5 μS/cm, electric fields frequency of 5 MHz), viable and nonviable yeast cell cells were separated as shown in (Fig. **16b**). This microfluidic chip was used to continuously separate the yeast cell mixture at the varying flow-rate in the range of 0.1-1 μl/min; thereby, resulting in the purity ranges of 95.9-97.3 and 64.5-74.3%, respectively, for the viable and nonviable yeast cells.

Hu *et al.* [113] furthered the idea of DEP-activated cell sorting in a continuous-flow manner, for isolating rare cells from complex mixtures in a glass/polyimide microfluidic chip. Cell mixtures of labeled bacteria and non-labeled bacteria were introduced in the DEP microfluidic sorter and were separated by controlling the electric fields. Rare target cells labeled by marker particles with different dielectric properties were efficiently separated from unlabeled cells. As shown in (Fig. **17a**), the buffer flow in the DEP chip was inverted such that the cell mixture flanked the buffer flow. As shown in (Fig. **17b**), the cell mixture was introduced at the side channels while the buffer solution of the same density and conductivity was pumped at the central inlet channel. Without an electric field, all DEP-responsive cells followed the streamlines and entered the waste channel. When the electrodes were energized, the labeled cells were selectively deflected from the sample stream into the buffer stream because of n-DEP deflection near the edges of the electrodes. Unlabeled cells weren't affected by the DEP force, leading to flowing along the sample solution and out at the wash channels. Thus only the labeled cells were collected effectively.

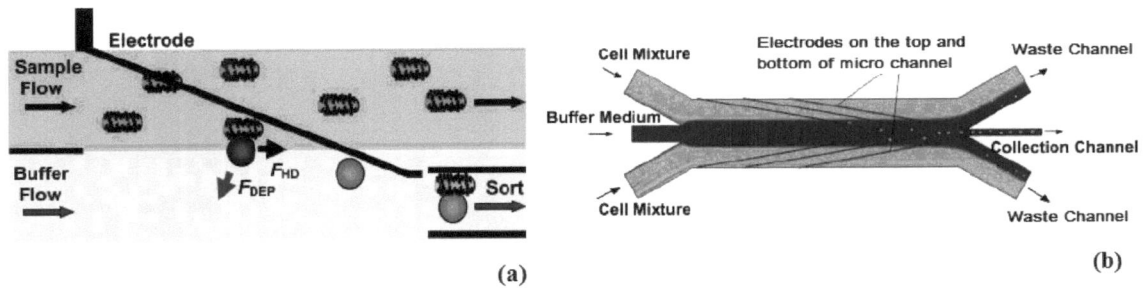

(a)   **(b)**

Figure 17: Operational principle of DEP-activated cell sorting [113]. **(a)** The DEP-activated cell sorting concept: Cells entering in the sample stream are only deflected into the collection stream if they are labeled with a dielectrophoretically responsive label. **(b)** Schematic view of the electrode region of the microchannels with sample and buffer inlets, as well as waste and collection outlets. Reprinted with permission from National Academy of Sciences, USA Copyright (2005).

From experiments, tagging cells allowed DEP sorting for rare target cells at rates up to 10,000 cells/s and with enrichment factors of more than 200. This DEP sorter offers the potential for automated, surface marker-specific cell sorting in a disposable format that is capable of simultaneously achieving high throughput, purity, and rare cell recovery.

Recently, a microfluidic chip based on a Printed Circuit Board (PCB) was designed and fabricated to separate and purify WBCs from erythrocytes by generating dielectrophoretic (DEP)-based cylinder-shaped cages [114].

Additionally, other research groups have revealed the DEP devices for RBC separation [115] and neuron collection [116]. A design of parallel and orthogonal electrodes microfabricated in a microfluidic device was used to test the motion of a suspension of erythrocytes response to a high-frequency Alternating Current (AC) field [115]. Under the optimal conditions (*e.g.*, 0.1 S/m isotonic phosphate buffer saline medium, 1 MHz of the electric field frequency), the largest cell mobility was observed. And the mobility was sensitive to RBC age and species, which can be used to separate different RBCs.

Tai *et al.* [116] reported a biochip capable of cell separation and nucleus collection. For the given chip, the DEP electrode array, microchannels, micropumps and microvalves were integrated on a microfluidic chip using MEMS technologies to perform cell transportation, separation and collection. Specific samples were separated by DEP electrodes and then collected by microvalves. Then the nucleus of the specific cells was collected after the cell lysis procedure.

Although it requires the fastidious property of low conductivity of the cell medium, combining inherent advantages of DEP with the recent microfabrication technology, the DEP technique is still considered as one of the most promising tools for cell separation from mixture samples [118-120] and many of them can be extended to cell separation from blood. The DEP method can permit to move, separate and position individual cells.

The group of Voldman [121, 122] developed a microdevice composed of a regular array of non-contact single-cell traps with proper electrode design. DEP was used to capture and manipulate single cells for monitoring the kinetics of fluorescent dye (calcein) uptake in HL-60 cells.

Cell trapping using DEP is based on the creation of energy traps of sizes comparable to the cells to be captured. Typically, traps created by positive DEP are used for separation of cells from flowing mixtures. Cells that are trapped are in stable equilibrium and can be released by simply turning off the electric field. The selectivity of cell capture from a mixture can be enhanced when the frequency of the electric field is chosen such that the unwanted cells are driven away by negative DEP. However, if the crossover frequency for the two types of cells to be separated is close, separation may not be very effective.

3.7. Hydrodynamic Separation

Hydrodynamic separation technique is the most simple and ideal principle to fractionate cells of interest. This method makes the cells separated depending on their cellular properties, such as size, shape, density and stiffness, which are determined by cellular behavior caused by interaction between the cells and surrounding medium or the gravitational force. These separation methods using only the geometries of microchannels and hydrodynamic forces are useful tools to resolve the problem of outer force field requirement in separation devices.

In 2004, the team of Yamada and Seki proposed a pinched-flow separation device to separate the cells according to their sizes [117, 123]. The concept of the pinched-flow separation is shown in (Fig. **18**), which is based on laminar flow behavior with negligible diffusion of the particles. A mixture of different-sized particles is injected into a solution, named a particle carrying buffer. The particle carrying buffer and another buffer without particles are forced to pass through a narrow, pinched channel segment, and then enter a widening channel. In the pinched segment, particles are aligned to one sidewall regardless of their sizes with the help of the buffer without particles which is pumped at a higher flow rate. At the boundary of the pinched and the broadened segments, particles are separated according to their sizes by the spreading flow profile, due to the fact that hydrodynamic forces act differently on particles, deflecting the small ones away from the big ones. When the fluid enters the widening channel, the laminar flow carries the two particle populations along different stream lines. The particles are thus separated across the width of the wide channel. If required, they can be collected further downstream.

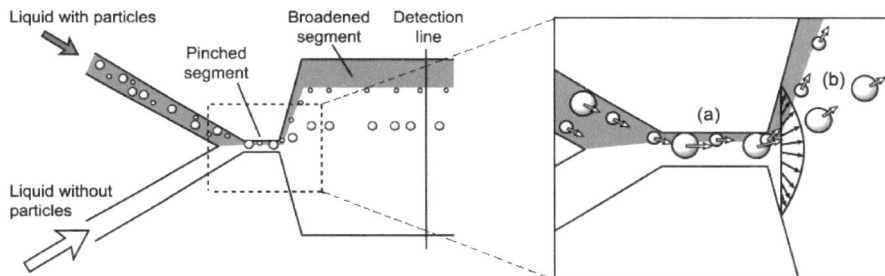

Figure 18: Principle of pinched flow fractionation to sort different-sized particles. The liquid containing particles is dark-colored [117]. Reprinted with permission from the American Chemical Society. Copyright (2004).

The separation of 15 µm and 30 µm polymer particles was used to test the pinched-flow separation devices. A microdevice with five branch channels connected to the pinched segment was designed to separate and collect different particles, shown in (Fig. **19a**). From experiments, large particles moved mainly toward outlet 2, while small particles moved toward outlet 1. As measured, 99.0% of small particles went to outlet 1, while the rest went to outlet 2. In the case of large particles, 91.6% went to outlet 2, while the rest went to outlet 3 [117].

In 2005, the same team of Yamada and Seki [124] improved the performance of pinched-flow separation devices. In the previous pinched-flow separation method (Fig. **19a**), the flow resistances of all branch channels were equal, so that the introduced liquid flow was uniformly distributed to all branch channels. In this case, even the largest particles whose diameters were comparable to the width of the pinched segment could go through the center branch channel, which means that branch 4 and 5 were useless.

In the asymmetric pinched-flow method (Fig **19b**), on the other hand, branch channels are arranged asymmetrically at the end of the pinched segment. One branch channel named the drain channel is made shorter and/or broader than the others (Branch 5 in Fig. **19b**) in order to reduce its flow resistance. Most of the carrier liquid is forced to leave *via* this outlet, whereas the remaining flow streams could fan out more over the remaining exits. In this case, the flow near Sidewall 1 in the pinched segment, which contains aligned particles, is effectively distributed to branch channels. Therefore, the difference in particle positions near Sidewall 1 can be effectively amplified. This device enabled the separation of a mixture of 1.0~5.0µm particles.

This device was also applied to the focusing of RBCs. Erythrocytes were successfully separated from blood. The blood-containing solution was pumped at Inlet 1 at 20 μL/h, while the buffer without blood was pumped at Inlet 2 was 1000μL/h. Since the diameter of erythrocytes is 7-8 μm, and the thickness is approximately 2 μm it was supposed that erythrocytes were aligned in the pinched segment as shown in (Fig. **19c**). The separation of cells whose shape is not spherical, the cell behavior is dominated by its minimum length. As measured, approximately 80% of erythrocytes went through Branch A3, while the rest went through Branch A2.

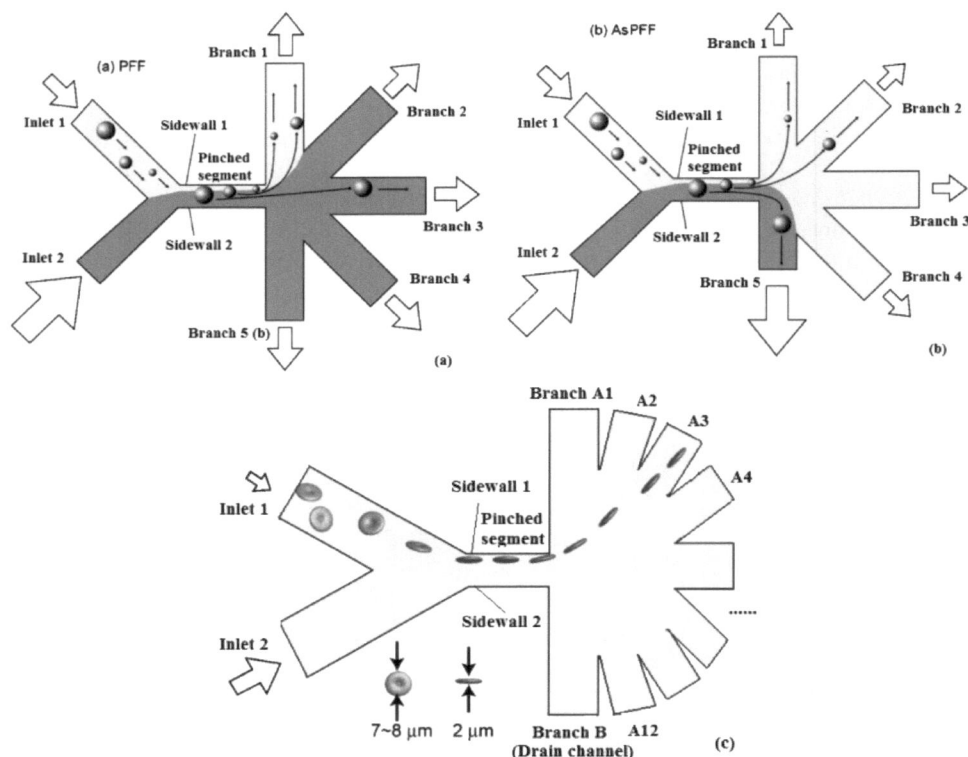

Figure 19: Principle of pinched flow fractionation to sort different-sized particles [124]. Liquid containing particles is light-colored, and liquid without particles is dark-colored. The size of an arrow represents the flow rate. **(a)** is normal pinched flow fractionation, where identical branch channels are arranged, and liquid flow in the pinched segment is uniformly distributed. **(b)** is asymmetric pinched flow fractionation, where one branch channel (drain channel) is designed to be short and/or broad, and liquid flow is asymmetrically distributed. **(c)** Schematic diagram of the erythrocyte alignment and separation in an asymmetric pinched flow fractionation. Reprinted with permission from the Royal Society of Chemistry. Copyright (2005).

Sai [125] and Zhang [126] also reported pinched-flow separation microfluidic devices. In the case of Sai, flow into the outlet channels was controlled by addition of a valve, which enabled the separation of 1 μm, 2 μm and 3 μm diameter particles from each other, as well as the separation of 0.5 μm and 0.8 μm particles. In the case of Zhang, a microdevice with three inlet channels leading into a bent and asymmetrically widening separation channel was designed. By varying the ratio of sample and buffer flow rates, microparticles could be forced onto a specific flow stream. Any differences in initial alignment were then enhanced by the flow profile within the bent channel. 10 μm and 25 μm particles were separated from each other.

The separation of pinched flow fractionation is based on particle size without labeling and doesn't have the problems of jamming or clogging. The dimensions of the pinched segment have implications for the applicable size range, and a significant amount of fine tuning has been necessary for the separation of sub-micron particles.

Blood circulation shows intriguing behaviors in microcapillaries, such as plasma skimming, erythrocytes selective movement and leukocyte margination. All the special behaviors can be used in microfluidic chips for plasma isolation and WBC separation in continuous flow. Based on plasma skimming, plasma with a low red blood cell concentration (hematocrit) can be continuously obtained [128]. When blood encounters a channel junction, erythrocytes selectively move into the branch with the faster flow rate. Therefore there is a higher concentration of RBCs in the branch with the faster flow rate, while there is a lower hematocrit in another branch with the lower flow rate This phenomenon of erythrocytes selective movement is referred to as "bifurcation law" or the "Zweifach-Fung effect" and has been demonstrated for plasma separation in continuous flow [129-131]. Both plasma skimming and "Zweifach-Fung effect" have been covered in in Chapter 2 in detail.

Base on leukocyte margination, auto-separation of leukocytes in the microfluidic device mimicking blood cell behavior in the vessels was reported by Shevkoplyas [127]. Collisions between RBCs that are predominantly found in the microchannel centre and WBCs lead to the white blood cells rolling along the channel wall. The phenomenon of leukocyte margination was used for leukocyte enrichment in continuous flow too.

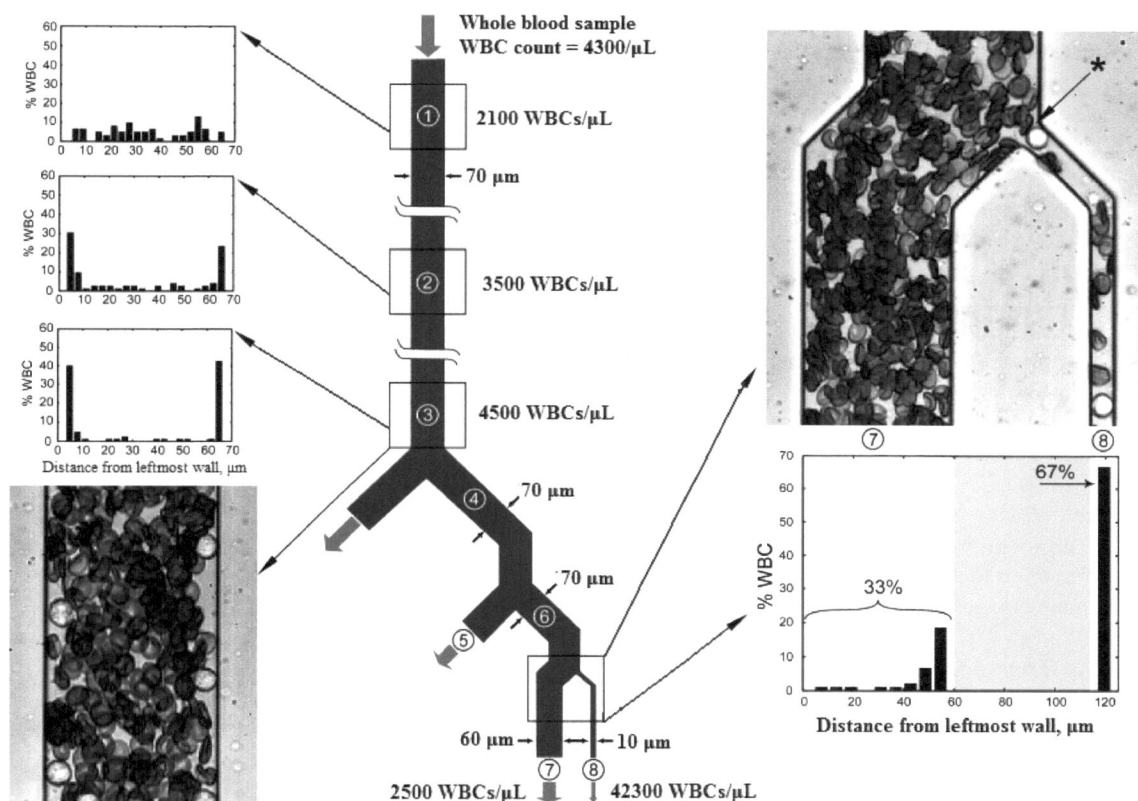

Figure 20: Microseparation device [127]. The geometry of the device is illustrated in the center; snapshots and plots of leukocyte (WBC) distributions are shown at the sides of the figure. Flow is from top to bottom in all segments. Length of the supply channel (points 1-3) is~5.5 mm; distance between the first and the third bifurcation is 0.48 mm. Flow rates in segments 7 and 8 were approximately 120 and 16 pL/s, respectively. Reprinted with permission from the American Chemical Society. Copyright (2005).

A microseparation device consisted of a simple network of microchannels schematically shown in (Fig. **20**). Whole blood entered the device through a 70μm wide supply channel by a small pressure gradient. All channels in the device are drained into a low-pressure reservoir. Initially, leukocytes are distributed uniformly across the channel. Because of plasma skimming, smaller, more flexible erythrocytes have the tendency to move to the center of the channel with the faster flow rate. Thus leukocytes were forced to move toward the sidewalls. Before the first bifurcation, the leukocyte distribution across the channel changed considerably, with most of the leukocytes traveling near the channel sidewalls (Fig. **20**, ③). As leukocytes moved to the sidewalls, they flowed slower than the rest of the blood, resulting in gradual accumulation of leukocytes to the sidewalls. And the concentration of leukocytes was increased from 2100 to 4500 WBCs/μL (Fig. **20**, ①-③). After passing through the first bifurcation, leukocytes continuously moved to the outside wall of each daughter channel. The asymmetry in leukocyte distribution in segment ④ caused the majority of the leukocytes to enter segment ⑥ of the second bifurcation, with very few entering segment ⑤. Approximately 67% of all leukocytes from the segment ⑥ entered the extraction channel ⑧, while the rest flowed along the right-hand sidewall of the larger channel ⑦ with a characteristic asymmetric distribution. Compared to the initial whole blood sample, the leukocyte concentration in the extraction channel ⑧ was increased ~10-fold. Although, RBCs remain present in the fraction of interest, the relative concentration of WBCs in relation to RBCs was increased by 34 times.

Another technique called deterministic lateral displacement has been applied to the size separation of particles and DNA molecules by pumping through an array of obstacles reported by Huang [132] and Inglis [133]. In this method, microposts are placed in rows within a microchannel. Each row of posts is shifted from the other by a distance which partly sets the critical separating size. The asymmetric bifurcation of laminar flow around obstacles lead particle to choose their path deterministically on the basis of their size. A small particle has a zigzag displacement path, whereas a large particle tends to flow straight.

After a number of rows, the particles can be collected separately. With a gap between posts of 1.6 μm it was possible to separate fluorescent particles with diameters of 0.6 μm, 0.8 μm and 1.0 μm as well as particles 0.7 μm and 0.9 μm in diameter [132]. This method is cost effective and can separate in parallel a large number of different-sized particles with a precision of up to 10 nm.

However, because of a high risk of clogging due to the high number of posts employed and the narrow gaps between them, Huang and coworkers [134] presented later an optimized device including additional regions alongside the active sorting arrays. These additional regions collect the larger particles that have been sorted. Although these regions also contain pillars to maintain the same pressure drop across the system, the gaps between the pillars are made larger to avoid clogging. This device was reported to remove all the particles larger than 1 μm from whole blood, allowing the collection of pure undiluted plasma. The latest technique is patented in [135].

The method was successfully tested for the separation of WBCs and RBCs [136]. A high-throughput and highly efficient microfluidic device was also developed for isolating rare Nucleated Red Blood Cells (NRBCs) from maternal blood based on the principle of deterministic lateral displacement and hemoglobin-based magnetic separation [137].

This paper reported a two-step enrichment process. In the first step, blood cell separation was based on cell size and their ability to deform. NRBCs are the only cells in whole blood that contain both a nucleus and hemoglobin. Note that RBCs and platelets are about 2 μm (thickness), whereas NRBCs and WBCs are larger than 5 to 10μm. Whole blood is passed through an array of microposts under laminar flow at a small angle relative to the array. Cells of RBCs and platelets that are small with respect to the flow stream drifted along the flow and stay in the stream. In contrast, cells of NRBCs and WBCs that are large compared to the flow stream are displaced out of the stream when they encounter a micropost. The microfluidic device effectively eliminates ~99.99% of RBCs in the first step.

In the second step, paramagnetic properties of NRBCs are used to remove WBCs without hemoglobin. The separated products from the first step are WBCs and NRBCs without most of RBCs. WBCs and NRBCs are passed through a magnetic column under a magnetic field, where NRBCs with hemoglobin are retained.

Finally, the deterministic lateral displacement microfluidic combined with magnetic separation has removed non-target red blood cells and white blood cells at a very high efficiency.

Li *et al.* developed a microfluidic device for continuous human blood cell subtype separation using the deterministic lateral displacement principle [138]. Even though all WBCs are spherical and have diameters within a narrow range (8 - 20 μm), the initial limitation for using this principle to separate WBC subtypes was solved by attaching larger polystyrene microbeads to one of the subtypes to amplify the size differences. CD4+ T helper lymphocytes were labeled with 25 μm polystyrene beads in a mixture of WBCs injected in a lateral displacement system. Up to 91% of the specific lymphocytes were therefore separated from the other kinds of WBCs. This article demonstrates the possibility of cell subtype sorting with the continuous-flow lateral displacement method.

With the approach of the deterministic lateral displacement, the pure population of one blood cell subtype can be effectively isolated. This method has the advantages of the simplicity, high speed and high resolution. Because many cells express unique surface markers, this method can theoretically be applied to separate any target cell type from a heterogeneous mixture for downstream analysis.

There are some other methods based on hydrodynamic separation. The papers from Carson [139, 140] described separation of WBCs in the microfabricated lattice structures using a model of activated sticking of cells with the wall.

Brody [141] designed and fabricated a novel microchip to separate cells based on the relative rigidity of the cell membrane, where cells were moved through a microarray *via* hydrodynamic flow.

In conclusion, these hydrodynamic separation techniques have the great advantage of not requiring any outer field. Therefore the manufacturing of these devices relies only on microchannel networks. Manufacturing of these channels can be achieved *via* hot-embossing or microinjection moulding. This holds the potential for cheap and mass-manufacturable devices, which is particularly relevant in the case of point-of-care devices. However fluidic-only separation devices, with the exception of the last example, can separate particles by size only.

4. CASE STUDY: MICRODEVICES FOR BLOOD CELL SEPARATION BASED ON THE CROSSFLOW FILTRATION PRINCIPLE

The fluidic flow based on dead-end filtration is perpendicular to the filtration barriers, so that when larger particles cannot pass through the barriers, they will build up near the filtration barriers, and then lead to colloid-cake formation, and finally result in clogging or jamming of channels. But in crossflow filtration, the flow direction of the fluid is parallel to the filtration structures. High tangential rates are utilized for transporting the permeated smaller particles from the feed stream to the lateral stream through the filtration structures, and the relative larger particles are still along the feed stream. Thereby crossflow filtration can dramatically restrict the problem of clogging or jamming.

In this case, pillar-type and weir-type crossflow filtration microchips were demonstrated for separation and collection of WBCs and RBCs [142]. Compared with dead-end filtration, crossflow filtration can effectively avoid the problem of clogging or jamming. In this example, the isolation efficiency of WBCs and the removing efficiency of RBCs were investigated. Additionally, an innovative crossflow chip with multilevel filtration barriers was designed and fabricated for the separation of plasma, WBCs and RBCs simultaneously with a great potential for blood sample pretreatment.

4.1. Design and Fabrication

The experiments of blood cell separation were conducted with silicon-based microfluidic chips fabricated mainly by photolithography and Deep Reactive Ion Etching (DRIE). Pillar-type and weir-type crossflow filtration chips were designed for separating RBCs from WBCs. Separation experiments were performed using anticoagulated whole rat blood. Video microscopy and blood counting chambers were used to evaluate the efficiency of blood cell separation.

It is well known that RBCs at rest assume a biconcave discoid shape with a diameter of ~8 μm and a thickness of ~2 μm and they are capable of passing through capillaries with less than half the diameter. Most of WBCs, mainly including 60%~75% neutrophils, 20%~45% lymphocytes and 2%~10% monocytes, are spherical with a diameter of more than 10 μm, though lymphocytes are small cells with 6~15 μm in diameter.

The designed crossflow filtration microfluidic chips consist of a poly (dimethylsiloxane) (PDMS)-glass compounded cover and a silicon substrate (Fig. **21**). The design of silicon substrate includes an inlet for introducing blood samples, filtration barriers within a tortuous channel, a WBC outlet for collecting WBCs, and a RBC outlet for collecting RBCs. The tortuous channel is divided into three sub-channels spaced by two-row filtration barriers in parallel. The silicon substrates were fabricated by using MEMS technologies.

Figure 21: Photograph of the silicon substrate.

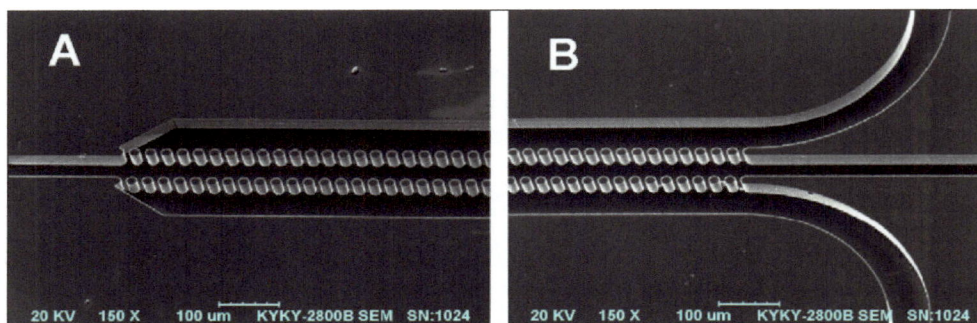

Figure 22: SEM micrographs of pillar-type barrier. SEM micrograph of the channel close to the inlet (**A**); SEM micrograph of the channel close to the outlet (**B**).

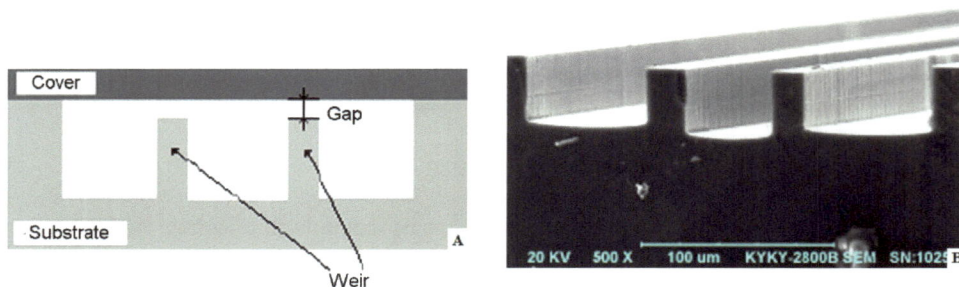

Figure 23: The design and construction of the weir-type barrier. The sketch map of the weir section (A); SEM micrograph of the weir.

The pillar-type filtration barrier is shown in (Fig. **22**), in which two-row pillars are about 20 μm in diameter and spaced by 6.5 μm. The weir-type filtration barrier is shown in (Fig. **23**), in which two integral weirs are 26.5 μm high and 20 μm wide. Both crossflow filtration microchips with different length of tortuous channels were designed to study the effect of channel length on the separation efficiency.

The patterned silicon substrates were sealed with a PDMS-glass compounded cover subsequently. The compounded cover, which was fabricated by thin-casting method, consists of a layer of PDMS and a layer of glass with one inlet and one outlet. The use of the glass-PDMS compounded cover can not only enhance the intensity of bonding but also help fix the tubing connections. PDMS with low surface free energy can be easily bonded with the silicon substrate and then peeled off with the help of ethanol.

4.2. Simulation

To study the flow patterns in all channels of crossflow filtration, flow dynamic simulations in a "straight" channel and a "tortuous" channel were performed by a commercial CFD software (Fluent) using the laminar model. All simulations were two-dimensional. The images of path lines of crossflow filtration are shown in (Fig. **24**). For the case of "straight" channel, the lateral flow of the solution occurs only in the regions close to the entrance of the channel, and becomes virtually stagnant in the regions far from the entrance. On the another hand, for the case of "tortuous" channel, the lateral flow occurs not only in the regions close to the entrance but also in the regions of the bend channel, producing higher tangential rates. As described above, high tangential rates could lead to the improvement in the separation efficiency. Thus the design of tortuous channel was chosen to separate blood cells in the following experiments.

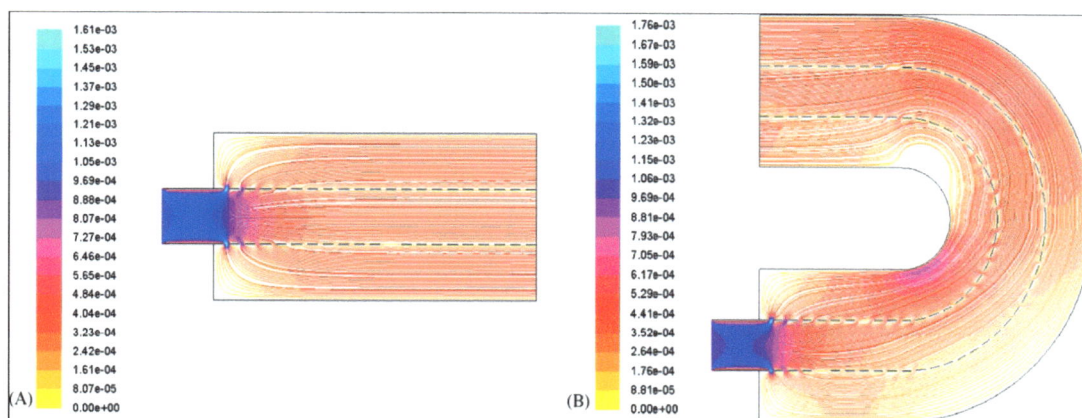

Figure 24: Images of path lines colored by velocity magnitude of a "straight" channel (**A**) and a "tortuous" channel (**B**).

4.3. Blood Cell Separation and Collection

Rat whole blood was collected from the orbital sinus into Vacutaner tubes containing EDTA-K2. Microchips were rinsed by 20 μL 0.9% NaCl solution prior to experiments. In a typical experiment based on crossflow filtration, the sample (2 μl anticoagulated rat whole blood, 18 μl 0.9%NaCl solution) was introduced into the microfluidic chip at the inlet by a peristaltic pump. Then the blood sample was divided into three branches by the separation barriers. These three branches were pumped out at two outlets and then collected into two 0.2 mL centrifugation tubes, respectively. Finally, the compounded cover was peeled off from the silicon substrate which enables device reusability after thoroughly rinse. RBCs and WBCs in raw whole blood samples and collected samples were counted by a blood counting chamber.

4.4. Experimental Results

Fabricated pillar-type and weir-type microfluidic chips based on crossflow filtration were evaluated by separating WBCs from RBCs. The size of filtration barriers should allow passage of RBCs oriented parallel to the main channel without any obstruction while block the free passage of the larger WBCs.

A schematic view and photograph of the microfluidic chip is illustrated in (Fig. **21**). The typical dimensions of the main channel are 30 μm×60 μm×160 mm (H×W×L). For the pillar-type chip, this main channel is connected to branch channels which are 30 μm×50 μm×160 mm (H×W×L) on both sides through sieves with micropillars which are about 20 μm in diameter and spaced by 6.5μm shown in (Fig. **22**). In this device, the filter gap is 30 μm high and 6.5 μm wide. For the weir-type chip, the filter gap is 3.5 μm high shown in (Fig. **23**). The geometry of the filter gap was chosen, aimed to guarantee that only the biconcave RBCs pass through the sieves into the branch channels and the larger spherical WBCs are contained in the main channel. Therefore if a cell encounters a barrier element and couldn't pass through it, it should be pushed along the main channel to the outlet due to the flow in the main channel, and hence there is no clogging or jamming. We confirmed the function of this chip in experiments using 200 μL solutions with more than 10^7 blood cells. These blood cells were pumped into crossflow filtration microchips with the gap of 3.5 μm. Under a constant pressure, the fluid rate was 4.6 μL/min at first, and was decreased to 3 μL/min after about 20 min, and then to 2.5 μL/min after about 1 hour. Though the fluid rate was noticed to decrease slowly, the microchip was never clogged or jammed during the whole experimental process.

4.4.1. Cell Concentration Effect

Diluted blood samples with different cell concentrations were used in separation experiments for both the pillar-type chip and the weir-type chip with a 160mm long channel at the flow rate of 10 μL/min. The experiments were carried out by introducing a diluted whole blood at the inlet port. WBCs not passing through filtration structures due to geometry restrictions were flowed out and collected at the first outlet port, whereas RBCs were filtered out through the filtration structures and collected at the second outlet port. WBCs and RBCs of the initial whole blood and the solutions collected at the outlet ports were counted by a blood counting chamber, respectively. Note that the original blood sample consists of 99% RBCs and only 1%WBCs. Therefore evaluating the removing efficiency of RBCs should be the first step for comprehensively assessing the performance of these microfluidic chips.

The effect of cell concentrations on the removing efficiency of RBCs is shown in (Fig. **25**) left. Removing efficiency of RBC was increased along with the increase of dilution times until the dilution ratio reached 50 times. When cell concentration was lower than 10^4/μL (diluted more than 50 times), the removing efficiency of RBCs became insensitive to cell concentration variations. Also shown in (Fig. **25**) left, the slope of RBC removing efficiency for the weir-type chip before saturation was much higher than that of the pillar-type counterpart. When dilution times was less than 20, RBC removing efficiency of the weir-type chip was much lower than that of the pillar-type chip since the larger gap in the pillar-type chip can remove much more red blood cells. However, when the blood sample was diluted more than 50 times, cell concentration was rather low (less than 10^4/μL). It is possible that RBCs which already passed through the barriers into the lateral streams may squeeze back into the feed stream by diffusion, which may limit further increase in the efficiency of RBC removal. When blood sample was diluted 100 times, RBC removing efficiency of the pillar-type chip was only 85.3% and that of the weir-type chip was 93.3%.

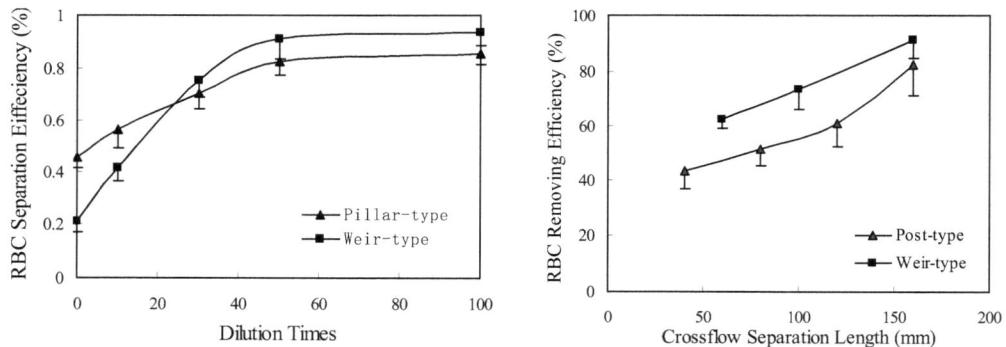

Figure 25: Effect of variation of cell concentration on the removing efficiency of RBC (left) and Effect of variation of separation length on the removing efficiency of RBC (right).

4.4.2. Effect of the Separation Channel Length on Separation Efficiency

Pillar-type and weir-type chips with different channel length were used to separate blood cells by loading whole blood samples with 10^4cells/μL at 10μL/min. The effect of variation of channel length on RBC removing efficiency is shown in (Fig. **25**) right. With the increase in the length of the separation channel, RBCs can be removed much more effectively. 82.3% RBCs were removed by the pillar-type chip with a separation length of 160 mm, while 91.2% RBCs by the weir-type counterpart. The weir-type chip showed a better performance due to the smaller filter gaps. Moreover, the collected solution at the first outlet port was introduced at the inlet port again to acquire WBCs with even higher purity. With twice separation, more than 95% RBCs were removed by using the weir-type chip.

Removing efficiencies of RBCs and WBCs are important factors to evaluate the performance of these chips. Only 8% WBCs was isolated from whole blood in the pillar-type chip with a 160 mm long channel, whereas 27.4% WBCs was isolated from whole blood in the weir-type device which was approximately twice higher as compared to the dead-end filtration microchip [16]. Besides the deformability of WBCs, larger gaps of the pillar-type filtration microstructures are the main reasons for lower isolation efficiency of WBCs. Therefore microchips with smaller gaps of filtration barriers should be used to achieve a higher isolation efficiency of WBCs. Note that sizes of filtration barriers need be large enough to enable RBCs to pass through the gaps.

After filtration experimentes, almost all blood cells were flowed out of the devices, with a few large WBCs retained in the chips (See (Fig. 26), which confirmed again that the problem of jamming or clogging didn't occur. Furthermore the crossflow filtration chips can be reused after washing, which is another advantage.

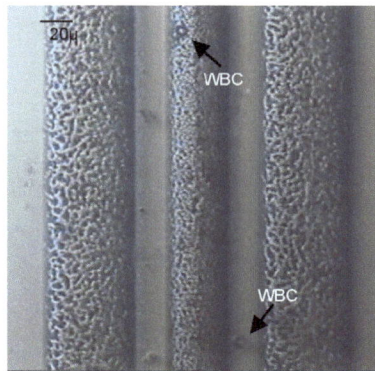

Figure 26: Photograph of weir-type microchip after filtration.

4.5. Multilevel Crossflow Filtration Microfluidic Chip

A novel crossflow filtration chip, which has four-row filtration barriers with increasing/decreasing filtration gaps, was proposed to separate three kinds of particles of different sizes. As shown in the (Fig. 27), additional micropillar arrays with a gap of 2.5 μm were designed and fabricated to separate plasma from blood cells. Plasma, RBCs and WBCs which were separated from the whole blood by this microfluidic chip with two-level filtration barriers can be collected respectively. In the collected plasma sample, there are only a few blood cells which indicated that the multi-level microchip was capable of separating and collecting different particles *via* various sizes in parallel.

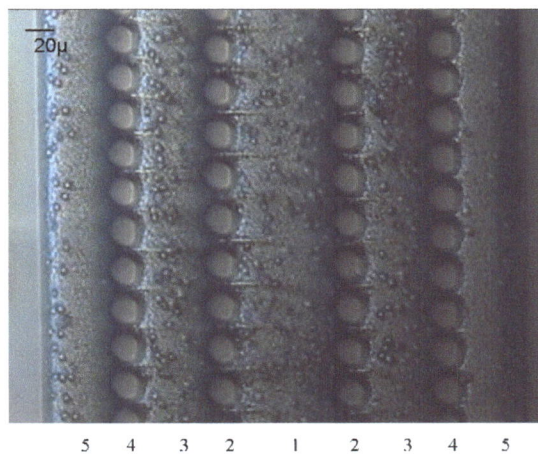

Figure 27: Photograph of multilevel crossflow filtration microchip for blood separation. The initial channel was divided into five channels by four rows of micropillars. 1: feed stream channel for WBC, 3: lateral stream channels for RBC, 5: lateral stream channels for blood plasma, 2: micropillars with a gap of 6μm, 4: micropillars with a gap of 3μm.

5. CONCLUSION

Initially complicated biologic samples, such as whole blood, should be directly pre-treated on the microchips without causing problems of clogging or jamming. For this specific purpose, size-filtering and collecting microchips based on crossflow filtration are shown to exhibit significant advantages over conventional methods and other microfluidic approaches. Crossflow filtration can strongly alleviate the problem of clogging. Moreover the chips based on crossflow filtration can provide advantages on: (i) fast separation time and collecting different cells at the same time; (ii) low volume of samples and reagents needed (low cost); and (iii) high reproducibility. Although particles in samples need be different in size, the crossflow filtration chips provide a feasible method for pre-treating raw whole blood samples.

ACKNOWLEDGEMENT

The authors greatly acknowledge the financial support from the National Science Foundation of China under the Grant numbers of 60701019 and 60501020.

REFERENCES

[1]   Prankerd T. The ageing of red cells. J Physiol 1958; 143 (2): 325-31.
[2]   Leif R, Vinograd J. The distribution of buoyant density of human erythrocytes in bovine albumin solutions. Proc Natl Acad Sci U S A 1964; 51 (3): 520-8.
[3]   Kimura E, Suzuki T, Kinoshita Y. Separation of reticulocytes by means of multi-layer centrifugation. Nature 1960; 188: 1201-2.
[4]   Danon D, Marikovsky V. Determination of density distribution of red cell population. J Lab Clin Med 1964; 64: 668-74.
[5]   Boyd E, Thomas D, Horton B, Huisman T. The quantities of various minor hemoglobin components in old and young human red blood cells. Clinica chimica acta 1967; 16 (3): 333-41.
[6]   Pertoft H, B ck O, Lindahl-Kiessling K. Separation of various blood cells in colloidal silica-polyvinylpyrrolidone gradients. Exp Cell Res 1968; 50 (2): 355-68.
[7]   Dooley D, Simpson J, Meryman H. Isolation of large numbers of fully viable human neutrophils: a preparative technique using percoll density gradient centrifugation. Exp Hematol 1982; 10 (7): 591-9.
[8]   Corash L, Piomelli S, Chen H, Seaman C, Gross E. Separation of erythrocytes according to age on a simplified density gradient. J Lab Clin Med 1974; 84 (1): 147-51.
[9]   Desimone J, Kleve L, Shaeffer J. Isolation of a reticulocyte-rich fraction from normal human blood on renografin gradients. J Lab Clin Med 1974; 84 (4): 517-24.
[10] Goebel K, Goebel F, Schubotz R, Schneider J. Red cell metabolic and membrane features in haemolytic anaemia of alcoholic liver disease (Zieve's syndrome). Br J Haematol 1977; 35 (4): 573-85.
[11] Vettore L, De Matteis MC, Zampini P. A new density gradient system for the separation of human red blood cells. Am J Hematol 1980; 8 (3): 291-7.
[12] Day R. Eosinophil cell separation from human peripheral blood. Immunol 1970; 18 (6): 955-9.
[13] Zborowski M, Fuh CB, Green R, *et al.* Immunomagnetic isolation of magnetoferritin-labeled cells in a modified ferrograph. Cytometry 1996; 24 (3): 251-9.
[14] Auditore-Hargreaves K, Heimfeld S, Berenson R. Selection and transplantation of hematopoietic stem and progenitor cells. Bioconjugate Chem; 1994, 5 (4): 287–300.
[15] Zborowski M, Sun L, Moore L, Williams P, Chalmers J. Continuous cell separation using novel magnetic quadrupole flow sorter. J Magn Magn Mater 1999; 194 (1): 224-30.
[16] Wilding P, Kricka LJ, Cheng J, Hvichia G, Shoffner MA, Fortina P. Integrated cell isolation and polymerase chain reaction analysis using silicon microfilter chambers. Anal Biochem 1998; 257 (2): 95-100.
[17] Zborowski M, Sun L, Moore LR, Stephen Williams P, Chalmers JJ. Continuous cell separation using novel magnetic quadrupole flow sorter. J Magn Magn Mater 1999; 194 (1): 224-30.
[18] Zborowski M, Ostera GR, Moore LR, Milliron S, Chalmers JJ, Schechter AN. Red blood cell magnetophoresis. Biophys J 2003; 84 (4): 2638-45.
[19] Takahashi M, Yoshino T, Takeyama H, Matsunaga T. Direct magnetic separation of immune cells from whole blood using bacterial magnetic particles displaying protein G. Biotechnol Prog 2009; 25 (1): 219-26.

[20] Wilding P, Pfahler J, Bau HH, Zemel JN, Kricka LJ. Manipulation and flow of biological fluids in straight channels micromachined in silicon. Clin Chem 1994; 40 (1): 43-7.

[21] Yang X, Yang JM, Tai YC, Ho CM. Micromachined membrane particle filters. Sensors and Actuators, A: Physical 1999; 73 (1-2): 184-91.

[22] Sethu P, Sin A, Toner M. Microfluidic diffusive filter for apheresis (leukapheresis). Lab Chip 2006; 6 (1): 83-9.

[23] Kim AS, Hoek EMV. Cake structure in dead-end membrane filtration: Monte Carlo simulations. Environ Eng Sci 2002; 19 (6): 373-86.

[24] Brody JP, Osborn TD, Forster FK, Yager P. A planar microfabricated fluid filter. Sensors and Actuators, A: Physical 1996; 54 (1-3): 704-8.

[25] Chu RW. Leukocytes in blood transfusion: adverse effects and their prevention. Hong Kong Med J 1999; 5 (3): 280-4.

[26] VanDelinder V, Groisman A. Perfusion in microfluidic cross-flow: Separation of white blood cells from whole blood and exchange of medium in a continuous flow. Anal Chem 2007; 79 (5): 2023-30.

[27] Metz S, Trautmann C, Bertsch A, Renaud P. Polyimide microfluidic devices with integrated nanoporous filtration areas manufactured by micromachining and ion track technology. J Micromech Microeng 2004; 14 (3): 324-31.

[28] Fu AY, Spence C, Scherer A, Arnold FH, Quake SR. A microfabricated fluorescence-activated cell sorter. Nat Biotechnol 1999; 17 (11): 1109-11.

[29] Fu AY, Chou HP, Spence C, Arnold FH, Quake SR. An integrated microfabricated cell sorter. Anal Chem 2002 ; 74 (11): 2451-7.

[30] Chen CC, Zappe S, Sahin O, *et al.* Design and operation of a microfluidic sorter for Drosophila embryos. Sensors and Actuators, B: Chemical 2004; 102 (1): 59-66.

[31] Krüger J, Singh K, O'Neill A, Jackson C, Morrison A, O'Brein P. Development of a microfluidic device for fluorescence activated cell sorting. J Micromech Microeng 2002; 12 (4): 486-94.

[32] Johann R, Renaud P. A simple mechanism for reliable particle sorting in a microdevice with combined electroosmotic and pressure-driven flow. Electrophoresis 2004; 25.

[33] Li PCH, Harrison DJ. Transport, manipulation, and reaction of biological cells on-chip using electrokinetic effects. Anal Chem 1997; 69 (8): 1564-6.

[34] McClain MA, Culbertson CT, Jacobson SC, Ramsey JM. Flow cytometry of Escherichia coli on microfluidic devices. Anal Chem 2001; 73 (21): 5334-8.

[35] Nieuwenhuis JH, Vellekoop MJ. Simulation study of dielectrophoretic particle sorters. Sensors and Actuators, B: Chemical 2004; 103 (1-2): 331-8.

[36] Wolff A, Perch-Nielsen IR, Larsen UD, Friis P, Goranovic G, Poulsen CR, *et al.* Integrating advanced functionality in a microfabricated high-throughput fluorescent-activated cell sorter. Lab Chip - Miniaturisation Chem Biol 2003; 3 (1): 22-7.

[37] Emmelkamp J, Wolbers F, Andersson H, *et al.* The potential of autofluorescence for the detection of single living cells for label-free cell sorting in microfluidic systems. Electrophoresis 2004; 25 (21-22): 3740-5.

[38] Wang MM, Tu E, Raymond DE, *et al.* Microfluidic sorting of mammalian cells by optical force switching. Nat Biotechnol 2005; 23 (1): 83-7.

[39] Ladavac K, Kasza K, Grier DG. Sorting mesoscopic objects with periodic potential landscapes: Optical fractionation. Phys Rev E Stat Nonlin Soft Matter Phys 2004; 70 (1 1).

[40] Ashkin A, Dziedzic JM, Bjorkholm JE, Chu S. Observation of a single-beam gradient force optical trap for dielectric particles. Opt Lett 1986; 11 (5): 288-90.

[41] MacDonald MP, Spalding GC, Dholakia K. Microfluidic sorting in an optical lattice. Nat 2003; 426 (6965): 421-4.

[42] Libál A, Reichhardt C, Jankó B, Reichhardt CJO. Dynamics, rectification, and fractionation for colloids on flashing substrates. Phys Rev Lett 2006; 96 (18): 188301-4.

[43] MacDonald MP, Neale S, Paterson L, Richies A, Dholakia K, Spalding GC. Cell cytometry with a light touch: Sorting microscopic matter with an optical lattice. J Biol Regul Homeost Agents 2004; 18 (2): 200-5.

[44] Smith RL, Spalding GC, Dholakia K, MacDonald MP. Colloidal sorting in dynamic optical lattices. J Opt A: Pure Appl Opt 2007; 9 (8).

[45] Chang-Hasnain CJ. Tunable VCSEL. IEEE J Sel Top Quant Electron 2000; 6 (6): 978-87.

[46] Deng T, Prentiss M, Whitesides GM. Fabrication of magnetic microfiltration systems using soft lithography. Appl Phys Lett 2002; 80 (3): 461.

[47]   Hartig R, Hausmann M, Schmitt J, Herrmann DBJ, Riedmiller M, Cremer C. Preparative continuous separation of biological particles by means of free-flow magnetophoresis in a free-flow electrophoresis chamber. Electrophoresis. 1992; 13 (9-10): 674-6.

[48]   McCloskey KE, Moore LR, Hoyos M, Rodriguez A, Chalmers JJ, Zborowski M. Magnetophoretic cell sorting is a function of antibody binding capacity. Biotechnol Prog 2003; 19 (3): 899-907.

[49]   Sun L, Zborowski M, Moore LR, Chalmers JJ. Continuous, flow-through immunomagnetic cell sorting in a quadrupole field. Cytometry 1998; 33 (4): 469-75.

[50]   Radisic M, Iyer RK, Murthy SK. Micro- and nanotechnology in cell separation. Int J Nanomedicine 2006; 1 (1): 3-14.

[51]   Chalmers JJ, Haam S, Zhao Y, *et al.* Quantification of cellular properties from external fields and resulting induced velocity: Magnetic susceptibility. Biotechnol Bioeng 1999; 64 (5): 519-26.

[52]   Reddy S, Moore LR, Sun L, Zborowski M, Chalmers JJ. Determination of the magnetic susceptibility of labeled particles by video imaging. Chem Eng Sci 1996; 51 (6): 947-56.

[53]   Chalmers JJ, Zborowski M, Moore L, Mandal S, Fang B, Sun L. Theoretical analysis of cell separation based on cell surface marker density. Biotechnol Bioeng 1998; 59 (1): 10-20.

[54]   McCloskey KE, Chalmers JJ, Zborowski M. Magnetophoretic mobilities correlate to antibody binding capacities. Cytometry 2000; 40 (4): 307-15.

[55]   McCloskey KE, Chalmers JJ, Zborowski M. Magnetic Cell Separation: Characterization of Magnetophoretic Mobility. Anal Chem 2003; 75 (24): 6868-74.

[56]   Nakamura M, Decker K, Chosy J, *et al.* Separation of a breast cancer cell line from human blood using a quadrupole magnetic flow sorter. Biotechnol Prog 2001; 17 (6): 1145-55.

[57]   Miltenyi S, Muller W, Weichel W, Radbruch A. High gradient magnetic cell separation with MACS. Cytometry 1990; 11 (2): 231-8.

[58]   Bu M, Christensen TB, Smistrup K, Wolff A, Hansen MF. Characterization of a microfluidic magnetic bead separator for high-throughput applications. Sensors and Actuators, A: Physical. 2008; 145-146 (1-2): 430-6.

[59]   Pamme N, Manz A. On-chip free-flow magnetophoresis: Continuous flow separation of magnetic particles and agglomerates. Anal Chem 2004; 76 (24): 7250-6.

[60]   Iliescu C, Barbarini E, Avram M, Xu G, Avram A. Microfluidic device for continuous magnetophoretic separation of red blood cells. 2008.

[61]   Iliescu C, Barbarini E, Avram M, Xu G, Avram A. Microfluidic device for continuous magnetophoretic separation of red blood cells. Symposium on Design, Test, Integration and Packaging of MEMS/MOEMS, Nice, France 2008.

[62]   Iliescu C, Xu G, Barbarini E, Avram M, Avram A. Microfluidic device for continuous magnetophoretic separation of white blood cells. Microsystem Technologies 2009; 15 (8): 1157-62.

[63]   Tanase M, Felton EJ, Gray DS, Hultgren A, Chen CS, Reich DH. Assembly of multicellular constructs and microarrays of cells using magnetic nanowires. Lab Chip - Miniaturisation Chem Biol 2005; 5 (6): 598-605.

[64]   Hultgren A, Tanase M, Felton EJ, *et al.* Optimization of yield in magnetic cell separations using nickel nanowires of different lengths. Biotechnol Prog 2005; 21 (2): 509-15.

[65]   Hultgren A, Tanase M, Chen CS, Reich DH. High-yield cell separations using magnetic nanowires. IEEE Trans Magn 2004; 40 (4 II): 2988-90.

[66]   Pamme N, Wilhelm C. Continuous sorting of magnetic cells *via* on-chip free-flow magnetophoresis. Lab Chip - Miniaturisation Chem Biol 2006; 6 (8): 974-80.

[67]   Pamme N, Eijkel JCT, Manz A. On-chip free-flow magnetophoresis: Separation and detection of mixtures of magnetic particles in continuous flow. J Magn Magn Mater 2006; 307 (2): 237-44.

[68]   Pamme N. Continuous flow separations in microfluidic devices. Lab Chip - Miniaturisation Chem Biol 2007; 7 (12): 1644-59.

[69]   Pamme N. Magnetism and microfluidics. Lab Chip - Miniaturisation Chem Biol 2006; 6 (1): 24-38.

[70]   Xia N, Hunt TP, Mayers BT, *et al.* Combined microfluidic-micromagnetic separation of living cells in continuous flow. Biomed Microdevices. 2006 Dec; 8 (4): 299-308.

[71]   Han KH, Frazier AB. Paramagnetic capture mode magnetophoretic microseparator for high efficiency blood cell separations. Lab on a Chip - Miniaturisation Chem Biol 2006; 6 (2): 265-73.

[72]   Han KH, Bruno Frazier A. A microfluidic system for continuous magnetophoretic separation of suspended cells using their native magnetic properties. 2005 NSTI Nanotechnology Conference and Trade Show, Anaheim, California, U.S.A. 2005.

[73]   Han KH, Bruno Frazier A. Continuous magnetophoretic separation of blood cells in microdevlce format. J Appl Phys 2004; 96 (10): 5797-802.

[74]   Blankenstein G, Larsen UD. Modular concept of a laboratory on a chip for chemical and biochemical analysis. Biosens Bioelectron 1998; 13 (3-4): 427-38.

[75]   Inglis DW, Riehn R, Austin RH, Sturm JC. Continuous microfluidic immunomagnetic cell separation. Appl Phys Lett 2004; 85 (21): 5093-5.

[76]   Inglis DW, Riehn R, Sturm JC, Austin RH. Microfluidic high gradient magnetic cell separation. J Appl Phys 2006; 99 (8).

[77]   Jung J, Han KH. Lateral-driven continuous magnetophoretic separation of blood cells. Appl Phys Lett 2008; 93 (22).

[78]   Chang WC, Lee LP, Liepmann D. Biomimetic technique for adhesion-based collection and separation of cells in a microfluidic channel. Lab Chip - Miniaturisation Chem Biol 2005; 5 (1): 64-73.

[79]   Seidl J, Knuechel R, Kunz-Schughart L, Inc M. Evaluation of membrane physiology following fluorescence activated or magnetic cell separation. Cytometry Part B: Clinical Cytometry. 1999;36 (2):102-11.

[80]   Revzin A, Sekine K, Sin A, Tompkins RG, Toner M. Development of a microfabricated cytometry platform for characterization and sorting of individual leukocytes. Lab Chip 2005; 5 (1): 30-7.

[81]   Nagrath S, Sequist LV, Maheswaran S, *et al.* Isolation of rare circulating tumour cells in cancer patients by microchip technology. Nature 2007 ; 450 (7173): 1235-U10.

[82]   Feng J. A Bio-MEMS Device for Separation of Breast Cancer Cells from Peripheral Whole Blood 2004.

[83]   Horsman KM, Barker SL, Ferrance JP, Forrest KA, Koen KA, Landers JP. Separation of sperm and epithelial cells in a microfabricated device: potential application to forensic analysis of sexual assault evidence. Anal Chem 2005 ; 77 (3): 742-9.

[84]   Coakley WT. Ultrasonic separations in analytical biotechnology. Trends Biotechnol 1997; 15 (12): 506-11.

[85]   Harris NR, Hill M, Beeby S, *et al.* A silicon microfluidic ultrasonic separator. Sensors and Actuators, B: Chemical 2003; 95 (1-3): 425-34.

[86]   Hawkes JJ, Barber RW, Emerson DR, Coakley WT. Continuous cell washing and mixing driven by an ultrasound standing wave within a microfluidic channel. Lab Chip - Miniaturisation Chem Biol 2004; 4 (5): 446-52.

[87]   Hawkes JJ, Coakley WT. Force field particle filter, combining ultrasound standing waves and laminar flow. Sensors and Actuators, B: Chemical 2001; 75 (3): 213-22.

[88]   Nilsson J, Evander M, Hammarström B, Laurell T. Review of cell and particle trapping in microfluidic systems. Anal Chim Acta 2009; 649 (2): 141-57.

[89]   Nilsson A, Petersson P, Laurell T. Whole blood plasmapheresis using acoustic separation chips. Proceedings of the Micrototal Analysis Systems 2006: 314-6.

[90]   Nilsson A, Petersson F, Jönsson H, Laurell T. Acoustic control of suspended particles in micro fluidic chips. Lab Chip - Miniaturisation Chem Biol 2004; 4 (2): 131-5.

[91]   Cousins C, Holownia P, Hawkes J, *et al.* Plasma preparation from whole blood using ultrasound. Ultrasound Med Biol 2000; 26 (5): 881.

[92]   Petersson F, Nilsson A, Holm C, Jönsson H, Laurell T. Separation of lipids from blood utilizing ultrasonic standing waves in microfluidic channels. Analyst 2004; 129 (10): 938-43.

[93]   Jönsson H, Nilsson A, Petersson F, Allers M, Laurell T. Particle separation using ultrasound can be used with human shed mediastinal blood. Perfusion 2005; 20 (1): 39-43.

[94]   Petersson F, Åberg L, Swärd-Nilsson AM, Laurell T. Free flow acoustophoresis: Microfluidic-based mode of particle and cell separation. Anal Chem 2007; 79 (14): 5117-23.

[95]   Petersson F, Nilsson A, Holm C, Jönsson H, Laurell T. Continuous separation of lipid particles from erythrocytes by means of laminar flow and acoustic standing wave forces. Lab on a Chip 2005; 5 (1): 20-2.

[96]   Petersson F, Nilsson A, Jönsson H, Laurell T. Carrier medium exchange through ultrasonic particle switching in microfluidic channels. Anal Chem 2005; 77 (5): 1216-21.

[97]   Doh I, Seo KS, Cho YH. A continuous cell separation chip using hydrodynamic dielectrophoresis process. Proceedings of the IEEE International Conference on Micro Electro Mechanical Systems (MEMS), Maastricht, Netherlands 2004.

[98]   Li Y, Kaler KVIS. Dielectrophoretic fluidic cell fractionation system. Anal Chim Acta 2004; 507 (1): 151-61.

[99]   Choi S, Park JK. Microfluidic system for dielectrophoretic separation based on a trapezoidal electrode array. Lab Chip - Miniaturisation Chem Biol 2005; 5 (10): 1161-7.

[100] Cui HH, Voldman J, He XF, Lim KM. Separation of particles by pulsed dielectrophoresis. Lab Chip 2009; 9 (16): 2306-12.

[101] Dürr M, Kentsch J, Müller T, Schnelle T, Stelzle M. Microdevices for manipulation and accumulation of micro- and nanoparticles by dielectrophoresis. Electrophoresis 2003; 24 (4): 722-31.

[102] Rousselet J, Markx GH, Pethig R. Separation of erythrocytes and latex beads by dielectrophoretic levitation and hyperlayer field-flow fractionation. Colloids and Surfaces A: Physicochemical and Engineering Aspects 1998; 140 (1-3): 209-16.

[103] Wang XB, Huang Y, Gascoyne PRC, Becker FF. Dielectrophoretic manipulation of particles. IEEE Trans Ind Appl 1997; 33 (3): 660-9.

[104] Yang J, Huang Y, Wang X, Wang XB, Becker FF, Gascoyne PRC. Dielectric properties of human leukocyte subpopulations determined by electrorotation as a cell separation criterion. Biophys J 1999; 76 (6): 3307-14.

[105] Gascoyne P, Mahidol C, Ruchirawat M, Satayavivad J, Watcharasit P, Becker FF. Microsample preparation by dielectrophoresis: isolation of malaria. Lab Chip 2002; 2 (2): 70-5.

[106] Wang XB, Yang J, Huang Y, Vykoukal J, Becker FF, Gascoyne PRC. Cell separation by dielectrophoretic field-flow-fractionation. Anal Chem 2000; 72 (4): 832-9.

[107] Yang J, Huang Y, Wang XB, Becker FF, Gascoyne PR. Cell separation on microfabricated electrodes using dielectrophoretic/gravitational field-flow fractionation. Anal Chem 1999; 71 (5): 911-8.

[108] Yang J, Huang Y, Wang XB, Becker FF, Gascoyne PRC. Differential analysis of human leukocytes by dielectrophoretic field- flow-fractionation. Biophys J 2000; 78 (5): 2680-9.

[109] Huang Y, Joo S, Duhon M, Heller M, Wallace B, Xu X. Dielectrophoretic cell separation and gene expression profiling on microelectronic chip arrays. Anal Chem 2002; 74 (14): 3362-71.

[110] Becker FF, Wang XB, Huang Y, Pethig R, Vykoukal J, Gascoyne PRC. Separation of human breast-cancer cells from blood by differential dielectric affinity. Proc Natl Acad Sci U S A 1995; 92 (3): 860-4.

[111] Doh I, Cho YH. A continuous cell separation chip using hydrodynamic dielectrophoresis (DEP) process. Sens Actuators A: Phys 2005; 121 (1): 59-65.

[112] Li HB, Bashir R. Dielectrophoretic separation and manipulation of live and heat-treated cells of Listeria on microfabricated devices with interdigitated electrodes. Sensor Actuat B-Chem. 2002; 86 (2-3): 215-21.

[113] Hu X, Bessette PH, Qian J, Meinhart CD, Daugherty PS, Soh HT. Marker-specific sorting of rare cells using dielectrophoresis. Proc Natl Acad Sci U S A 2005; 102 (44): 15757-61.

[114] Borgatti M, Altomare L, Baruffa M, Fabbri E, Breveglieri G, Feriotto G, *et al.* Separation of white blood cells from erythrocytes on a dielectrophoresis (DEP) based 'Lab-on-a-chip' device. Int J Mol Med 2005; 15 (6): 913-20.

[115] Minerick AR, Zhou R, Takhistov P, Chang HC. Manipulation and characterization of red blood cells with alternating current fields in microdevices. Electrophoresis 2003; 24 (21): 3703-17.

[116] Tai CH, Hsiung SK, Chen CY, Tsai ML, Lee GB. Automatic microfluidic platform for cell separation and nucleus collection. Biomed Microdevices 2007; 9 (4): 533-43.

[117] Yamada M, Nakashima M, Seki M. Pinched flow fractionation: Continuous size separation of particles utilizing a laminar flow profile in a pinched microchannel. Anal Chem 2004; 76 (18): 5465-71.

[118] Gascoyne P, Vykoukal J. Dielectrophoresis-based sample handling in general-purpose programmable diagnostic instruments. Proc IEEE Inst Electr Electron Eng 2004; 92 (1): 22-42.

[119] Gascoyne PR, Vykoukal J. Particle separation by dielectrophoresis. Electrophoresis. 2002; 23 (13): 1973-83.

[120] Hughes MP. Strategies for dielectrophoretic separation in laboratory-on-a-chip systems. Electrophoresis 2002; 23 (16): 2569-82.

[121] Voldman J, Gray ML, Toner M, Schmidt MA. A microfabrication-based dynamic array cytometer. Anal Chem 2002; 74 (16): 3984-90.

[122] Rosenthal A, Taff BM, Voldman J. Quantitative modeling of dielectrophoretic traps. Lab Chip 2006; 6 (4): 508-15.

[123] Yamada M, Kasim V, Nakashima M, Edahiro J, Seki M. Continuous cell partitioning using an aqueous two-phase flow system in microfluidic devices. Biotechnol Bioeng 2004; 88 (4): 489-94.

[124] Takagi J, Yamada M, Yasuda M, Seki M. Continuous particle separation in a microchannel having asymmetrically arranged multiple branches. Lab Chip 2005; 5 (7): 778-84.

[125] Sai Y, Yamada M, Yasuda M, Seki M. Continuous separation of particles using a microfluidic device equipped with flow rate control valves. J Chromatogr A 2006; 1127 (1-2): 214-20.

[126] Zhang XL, Cooper JM, Monaghan PB, Haswell SJ. Continuous flow separation of particles within an asymmetric microfluidic device. Lab Chip 2006; 6 (4): 561-6.

[127] Shevkoplyas SS, Yoshida T, Munn LL, Bitensky MW. Biomimetic autoseparation of leukocytes from whole blood in a microfluidic device. Anal Chem 2005; 77 (3): 933-7.

[128] Jäggi RD, Sandoz R, Effenhauser CS. Microfluidic depletion of red blood cells from whole blood in high-aspect-ratio microchannels. Microfluid Nanofluidics 2007; 3 (1): 47-53.

[129] Yang S, Ündar A, Zahn JD. A microfluidic device for continuous, real time blood plasma separation. Lab Chip - Miniaturisation Chem Biol 2006; 6 (7): 871-80.

[130] Yang S, Ündar A, Zahn JD. Blood plasma separation in microfluidic channels using flow rate control. ASAIO J 2005; 51 (5): 585-90.

[131] Yang S, Ündar A, Zahn JD. Biological fluid separation in microfluidic channels using flow rate control. ASME International Mechanical Engineering Congress and Exposition, Orlando, FL, U.S.A. 2005.

[132] Huang LR, Cox EC, Austin RH, Sturm JC. Continuous particle separation through deterministic lateral displacement. Science 2004; 304 (5673): 987-90.

[133] Inglis DW, Davis JA, Austin RH, Sturm JC. Critical particle size for fractionation by deterministic lateral displacement. Lab Chip - Miniaturisation Chem Biol 2006; 6 (5): 655-8.

[134] Davis J, Inglis D, Morton K, *et al.* Deterministic hydrodynamics: Taking blood apart. P Proc Natl Acad Sci U S A 2006; 103 (40): 14779.

[135] Huang L, Barber T, Carvalho B, *et al.* Devices and methods for enrichment and alteration of cells and other particles. United States Patent 2007; 20070026381.

[136] Zheng S, Tai YC, Kasdan H, ed. A micro device for separation of erythrocytes and leukocytes in human blood. Conf Proc IEEE Eng Med Biol Soc 2005;1:1024-7.

[137] Huang R, Barber TA, Schmidt MA, *et al.* A microfluidics approach for the isolation of nucleated red blood cells (NRBCs) from the peripheral blood of pregnant women. Prenat Diagn 2008; 28 (10): 892-9.

[138] Li N, Kamei DT, Ho CM. On-chip continuous blood cell subtype separation by deterministic lateral displacement. 2nd IEEE International Conference on Nano/Micro Engineered and Molecular Systems, Bangkok, Thailand 2007.

[139] Carlson R, Gabel C, Chan S, Austin R. Activation and sorting of human white blood cells. Biomed Microdevices 1998; 1 (1): 39-47.

[140] Carlson RH, Gabel CV, Chan SS, Austin RH, Brody JP, Winkelman JW. Self-sorting of white blood cells in a lattice. Phys Rev Lett 1997; 79 (11): 2149-52.

[141] Brody JP, Han Y, Austin RH, Bitensky M. Deformation and flow of red blood cells in a synthetic lattice: Evidence for an active cytoskeleton. Biophys J 1995; 68 (6): 2224-32.

[142] Chen X, Cui DF, Liu CC, Li H. Microfluidic chip for blood cell separation and collection based on crossflow filtration. Sens Actuat B 2008; 130 (1): 216-21.

CHAPTER 4

Microfluidic Devices Targeting Blood Cell Lysis

Xing Chen[*], Dafu Cui and Jian Chen

State Key Laboratory of Transducer Technology, Institute of Electronics, Chinese Academy of Sciences, Beijing, China

Abstract: Cell lysis is an essential step in whole blood sample preparation, which can pave road for intracellular proteomic analysis or genomic analysis and therefore cell membrane rupture is requested for protein and gene release. In this chapter, we firstly introduce reported microfluidic devices for cell lysis. And then an example using the chemical cell lysis method in microfluidic chip is put forward in which a sandwich microfluidic chip was designed, fabricated and tested to disrupt blood cell membranes.

Keywords: Cell lysis, MEMS, microfluidics.

1. INTRODUCTION

The initial step in immunoassays for clinic diagnostics and genomic analysis is the release of the intracellular molecules such as proteins, biomarkers, and gene from the cells of interest, or cell lysis. Typical laboratory methodologies for off-chip cell lysis include the use of enzymes (lysozyme [1]), chemical lytic agents (detergents, alkaline buffers) [2], and mechanical forces (sonication [3], bead milling [4]). However, the traditional methods suffer from the problems such as high sample and reagent usage, high time and manual labor consumption, *etc.*

With the development of Micro Total Analysis Systems (μTAS), a few groups have been pushing forward the development of miniaturized devices for cell lysis. Methods for cell lysis in microfluidic devices can be divided into mechanical lysis and non-mechanical counterparts. Mechanical lysis is realized by nanostructures with sharp corners [5] or ultrasonic forces [6, 7]. The methods of non-mechanical lysis mainly include thermal lysis [8-10] , electronic-based lysis [11-13] and chemical-based lysis [14, 15]. Compared to macro counterparts, microfluidic tools have significant advantages on lower sample consumption and higher throughput.

In this chapter, we first summarize the traditional methods for cell rapture, and then review the state-of-art miniaturized methods for cell lysis. Finally we propose a case example to illuminate how to design, fabricate and test microdevices for cell lysis. In this example, a network of microfluidic channels was designed with numerical simulation confirmation. The fabrication process of the microdevice was investigated and optimized. The device performance was characterized and tested by blood cell lysis experiments.

2. CURRENT MICRODEVICES FOR CELL LYSIS

On-chip cell lysis has been implemented by several groups using different methods, which can be categorized into six major groups. Mechanical lysis employs cellular contact forces to crush or burst cells, which can be further divided into mechanical force lysis based on sharp nanostructures or the sonication. Thermal lysis uses high temperatures to disrupt the cell membrane. Chemical lysis uses chemical buffers or enzymes to break down cell membranes. Electrical lysis induces cell membrane porosity with a low-strength electric field or realizes a complete cell lysis with a high-strength electric field.

In this chapter, the performance of these techniques is discussed and compared. In addition, relevant fabrication methods are explained to help readers understand the microfluidics based cell lysis and their

Address correspondence to Xing Chen: State Key Laboratory of Transducer Technology, Institute of Electronics, Chinese Academy of Sciences, Beijing 100190, China; Phone and fax: +86-10-58887188; E-mail: xchen@mail.ie.ac.cn

potential integration with other whole blood pretreatment components.

2.1. Mechanical Lysis with Microstructures or Microbeads

Cell lysis occurs when the cell membrane (and/or cell wall) is disrupted in some way allowing the contents of the cytoplasm to be released. Perhaps the most conceptually simple method to achieve lysis is to use a mechanical force to tear or puncture the membrane. It is known that subjecting cells to shear and frictional forces is used to rupture cell membranes.

Adding sharp nanostructures to contraction sidewalls and increasing volumetric flow rates were implemented in a microfluidic chip to concentrate and amplify frictional forces and therefore penetrate the cell membrane, which was reported by Carlo [5].

In this device, nanostructural barbs were designed and fabricated on a silicon wafer using MEMS techniques including a two-mask bulk etching and the Pyrex wafer bonding process. First, a silicon oxide layer was grown on an n-type silicon wafer using wet thermal oxidation. This oxide was patterned and etched with the designed microfluidic channel, filter, and collection geometries. Photoresist was then spun on and patterned with fluid access ports. A Deep Reactive Ion Etch (DRIE) was then performed to etch through the wafers to form the patterned high aspect ratio backside access ports. Next the photoresist layer was stripped and channel and filter geometries were etched into the wafer to a depth of 20 μm using the silicon oxide mask and a modified DRIE process. This etching step formed the sharp nanostructure for mechanical lysis, which the authors termed as nano-knives (Fig. **1**), the winding biomolecule collection chamber, and connecting microchannels. Next, the wafer was cleaned in piranha to remove any polymer residue left from the DRIE process. Any the remaining silicon dioxide was then removed from the wafer by washing in HF. Immediately after the HF washing, a thin thermal expansion coefficient matched Pyrex wafer was anodically bonded to the bulk-etched silicon wafer forming enclosed channels with backside access ports in the silicon.

In this fabrication process, the nano-knives were formed using a modified DRIE recipe that increases the "scalloping" effect. In the modified recipe the etch step was more isotropic, enhancing the scalloped effect. Using this modified recipe, the effect of sharp nano-knife tips were formed at corners as two orthogonal scallops met (Fig. **1**). The distance between tips in the upper region of the pillar was characterized as 0.34 ± 0.03 μm from SEM images, while the radius of curvature at the tips was less than 25 nm. A schematic diagram of the nano-knives is shown in (Fig. **1c)**. These nano-structural barbs were less than 25 nm, and were used to disrupt sheep blood cells, and then the intracellular materials were released.

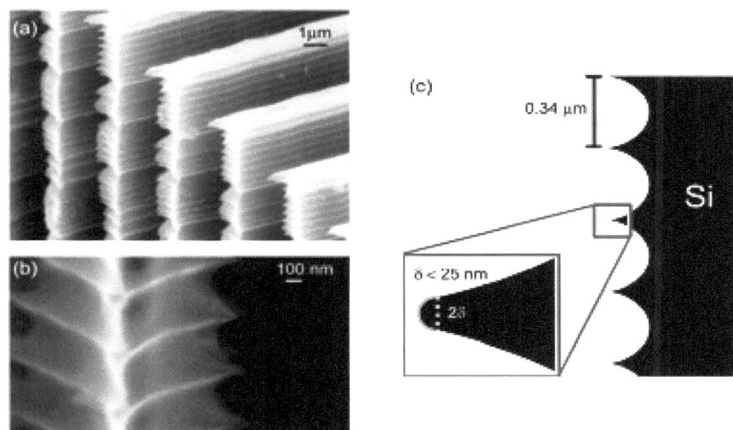

Figure 1: (**a-b**) SEM of the nano-knives are shown. Sharp protrusions are clearly seen as orthogonal scallops meet at corners during the DRIE process. (**c**) A schematic drawing of the nano-knives is shown describing the geometry. Distance between barbs is ~ 0.34μm and radius of curvature of tips is below 25 nm [5]. Reprinted with permission from the Royal Society of Chemistry. Copyright (2003).

Since cells' membrane could be burst because of cell deformation [16], microchannels shaped with a curved cross section on one side and a flat membrane on the other side were used to rush cells and break their membranes based on pressure forces [17]. In this design, the flat membrane is used to separate the fluid channel from a pneumatic control channel. The membrane is forced to expand into the fluid channel and crush any cells trapped there by the pressure control. As the pressure on the cells is slowly increased from ambient, the cells flatten out when they are compressed and deformed. In experiments, as the pressure approached 25 kPa in the system, the cells began to split open and spilt their internal cellular structures. Complete lysis was recorded when the pressure approached 30 kPa.

In another method based on mechanical lysis, spherical particles (microbeads) in a lysis chamber were used to disrupt cells. The mechanism of mechanical lysis is to generate confined periodic flow fields with angular accelerations that provide strong vortical flow fields and large shear rates between beads in the liquid solution.

This purely mechanical lysis method was proposed based on a rapid granular shear flow [18, 19] in an annular chamber of a CD which spins about a horizontal axis of rotation alternating between the forward and reverse directions. When the CD is at rest, spherical particles, such as glass beads, in an aqueous medium containing cells to be disrupted, lie in a pool at the bottom of the chamber. However, during spinning all particles are dragged up by the surrounding medium and uniformly coat the outer wall of the chamber in response to both shear and centrifugal forces. This flow is often referred to as "rimming flow". In order for a cell to be disrupted in a rapid granular flow, the cell should be brought into physical contact with the colliding particles. The bead interactions with the cells consist of two main types: impulsive contact (collision) due to the beads responding to the centrifugal field and sustained contact (friction) due to the shearing.

Then Kim [20] reported a microfluidic CD platform capable of causing disruption of mammalian (CHO-K1), bacterial (*Escherichia coli*), and yeast (*Saccharomyces cerevisiae*) cells. In this design, interactions between beads and cells are generated in the rimming flow established inside a partially filled annular chamber in the CD rotating around a horizontal axis, used to rupture cells.

As to the fabrication, for the purpose of mechanical cell disruption, an ultra-thick SU-8 fabrication process was developed to fabricate a mold featuring extra high structures (~1 mm) so that sufficiently high lysing chambers could be formed in the PDMS for the interaction of beads with cells.

A thick layer of SU-8 100 was spin coated onto a Si wafer. After these processes of prebaking, postbaking, photolithography, and development, the mold master was prepared. Then the PDMS precursor and curing agent were mixed thoroughly in a weight ratio of 10:1. After degassing, it was poured and cured on the SU-8 master mold which had a rim fashioned around the substrate to contain the liquid PDMS.

The rotating CD platform was assembled by sandwiching the micromachined PDMS sheet between two polycarbonate discs (Fig. **2**). In (Fig. **2a**), a CD with a typical annulus chamber was designed for cell lysis. The wavy inner wall (see (Fig. **2b**) of the chamber was intended to assist in lifting up a solid-liquid suspension at the moment of spinning the CD (*i.e.*, little boundary layer growing). In this design, advantages of the particle interactions were taken since many particles stayed longer around the inner wall.

Cell disruption efficiency was verified either through direct microscopic viewing or measurement of the DNA concentration after cell lysing. Lysis efficiency relative to a conventional lysis protocol was approximately 65%. In the long term, this work was geared towards CD based sample-to-answer nucleic acid analysis which could include cell lysis, DNA purification, DNA amplification, and DNA hybridization detection.

Various means of mechanical lysis have been developed as commercially available instruments. Most of the protocols that have been developed for hard-to-lyse samples are not suitable for routine automated use in biodefense and other applications [7, 21].

Two mechanical lysis miniaturized devices were developed as compact, robust components to provide rapid sample preparation for nucleic acid diagnostic systems [22].

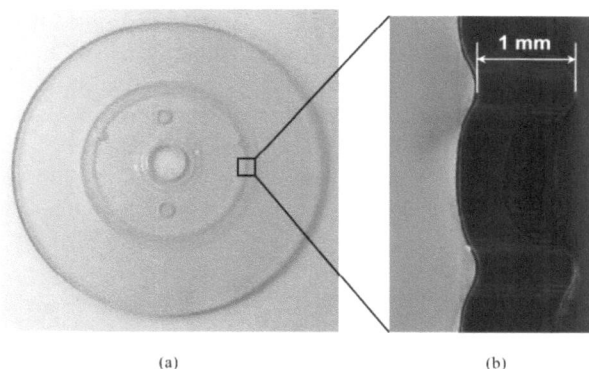

Figure 2: The CD designed for cell lysing [20]. **(a)** CD with an annular chamber (total volume of 1 mL) **(b)** SEM photo of a wavy inner wall of the chamber. Reprinted with permission from the Royal Society of Chemistry. Copyright (2004).

The first mechanical lysis platform was a Micro Bead Beater, which was capable of ultrarapid lysis (>90% lysis in 30 s) of micro volumes (<80 μL) of Bacillus spores in a continuous-flow format or in a disposable single-tube format. This Micro Bead Beater was also capable of processing much larger volumes of solutions containing spores or vegetative cells using a continuous flow mode. A second mechanical lysis device was a Microfluidic Bead Blender, which was designed as a disposable component. A small electric motor was used to spin vanes within the bead-laden solution in the second mechanical lysis platform.

The Micro Bead Beater and the Microfluidic Bead Blender were characterized using a Pico-Green dsDNA-binding fluorescence dye assay and a real time PCR assay. Results consistently showed rapid lysis (<1 min) in small volume (~50-100 μL) samples using these devices.

In summary, the method of mechanical lysis is a simple and useful technique, which can form physical contact with cells and therefore lead to cell membrane rupture. This method can provide non-adulterated cell lysates without considering the issue of lysate contamination since no detergents are involved in the cell lysis procedure. However, fabricating sharp nanostructures and controlling of the interactions between mcirobeads and cells are challenging, which limits the further application of this technique.

2.2. Sonication Lysis

Sonication based cell lysis is another method of mechanical cell lysis that differs from the physical contact methods. This technique is a widely practiced method to rapidly disrupt a variety of cell types [24-31]. The mechanism in which sonication forces disrupt cells has been proposed to be gaseous cavitation by employing ultrasonic agitation to create pressure waves with enough energy. In this process, pockets of air form from the dissolved gases in a solution and then rapidly collapse to a portion of the original size, creating high pressure and temperature microenvironments that are damaging to cells.

The group of Northrup [6, 7] integrated sonication into a microfluidic system and showed that increasing the fluid pressure enhanced the coupling between the ultrasonic horn tip and the liquid region, which lead to more efficient cell lysis. Cell disruptions were implemented by using ultrasonic energy transmitted through a flexible interface into a liquid region. A fluidic chamber was in contact with the ultrasonic horn through a flexible interface.

This miniaturized cell lysis device with a piezoelectric film was used to disrupt bacterial spores in 30 sec [6]. After being captured on a filter and washed with water, the bacterial spores were disrupted at the presence of glass beads by applying ultrasonic energy through a thin-film flexible interface. This device incorporated a glass bead suspension in the sample and used an external ultrasonic transducer to generate ultrasonic waves in the solution that were powerful enough to rupture even bacterial spores.

Then a piezoelectric microfluidic minisonicator was developed to disrupt Eukaryotic human leukemia HL-60 cells and *Bacillus subtilis* bacterial spores [23]. A microfluidic channel with integrated transducers was designed and fabricated. The geometry of the device was shown in (Fig. **3**). The device is composed of two parts, the channel and the transducers. The channel was fabricated on a glass substrate, whereas the transducers were fabricated on a quartz substrate. The fabricated process of the channel was as follows. First the glass was coated with polysilicon as a masking layer. Then the glass was patterned by the photolithography technology. Finally the channel was formed by wet-etching the glass substrate in a 50:1 hydrofluoric acid solution.

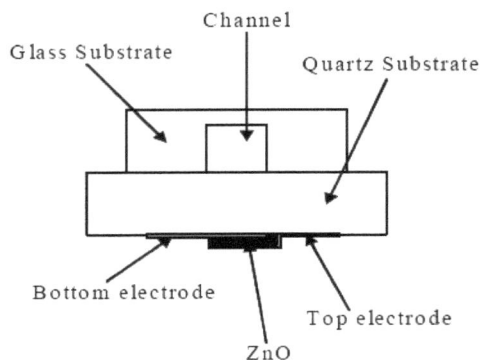

Figure 3: Schematic cross-section of the microfluidic channel [23]. Reprinted with permission from Elsevier. Copyright (2005).

The piezoelectric transducers used in the design were integrated onto the channel floor by depositing a layer of zinc oxide between two layers of gold on a quartz substrate. Quartz was chosen as a substrate material because of its low loss coefficient for acoustic waves. The fabrication steps of the transducer were as follows. First, a 0.1-µm film of gold was sputtered and patterned over a quartz substrate to serve as one of the electrodes of the transducer. This step was followed by the deposition of 8 µm of zinc oxide using a shadow mask. Finally, another 0.1 µm thick layer of gold was deposited and patterned using the lift-off technology to serve as the top electrode of the transducer. Finally, the device was assembled by gluing the channel onto the opposite side of the substrate where the transducers were immobilized.

Using this microfabricated device, 80% HL-60 cells were lyzed after 3 sec of sonication and 50% *Bacillus subtilis* bacterial spores were disrupted within 30 sec of sonication.

The method of sonication has advantages such as simple fabrication processes and high lysis efficiency, which are usually used for lyzing cells of microorganism. However, it is difficult to integrate with other microfluidic components since it can not be further miniaturized.

2.3. Thermal Lysis

Thermal lysis of cells is one of the most well established lysis techniques, and is common in conventional laboratory settings. At high temperatures, proteins within the cell membranes are denatured, irreparably damaging the cell and releasing the cytoplasmic contents. This thermal damage is one of the most common traditional methods by immersing a sample tube in a boiling water bath.

This lysis method is now being used more due to easy integration with PCR-based microfluidic systems. For instance, literatures *et al.* [9, 10] reported a PCR microchip that could thermally lyze cells using the heater of the PCR reactor.

The steps of cell lysis, multiplex PCR amplification, and electrophoretic analysis were implemented sequentially on a monolithic microchip device [10]. The entire microchip was mounted in a commercial thermal cycler equipped with a slide griddle and a hot bonnet. The entire microchip was thermally cycled to lyse cells and to amplify DNA. Whole *E. coli* cells were thermally lysed during the first 4 min of the initial temperature setting of 94°C.

A fully integrated biochip that consists of a mixing unit for rare cell capture using immunomagnetic separation, a cell preconcentration/purification/lysis/PCR unit, and a DNA microarray chamber was also developed to perform DNA analysis of complex biological sample solutions [32]. The chamber to capture and preconcentrate target cells was also used for subsequent cell lysis and PCR. Pathogenic bacteria was captured from a whole blood sample and then thermally disrupted in the first PCR thermal cycle by using embedded resistive heaters. This system was effective for single samples, but cannot be used as part of a continuous flow assay or with downstream analysis techniques that require a sample free of protein and other cellular contaminants.

A microfluidic chip with a cell lysis reactor, a micromixer, a PCR chamber, a sample driving module and an on-chip temperature control system was reported by Lee *et al.* [33]. The cell lysis was performed in a micro cell lysis reactor by using a thin-film heater at 95 °C.

Yeung *et al.* further developed the method of thermal lysis by using the PCR heater [8]. A single silicon and glass-based microchamber was used for thermal lysis, DNA amplification, and electrochemical detection of the target DNA, shown in (Fig. **4**). A thin-film heater and a temperature sensor were patterned on the silicon substrate. An array of Indium Tin Oxide (ITO) electrodes was constructed within the microchamber as the transduction element. The chamber was maintained at 90 °C for 5 min to lyse the cell and at the same time denature the genomic DNA. This process again showed effective use of thermal lysis in a microfluidic device, but its complicated operation steps and inability to function as a continuous flow device gave it limited use in µTAS applications.

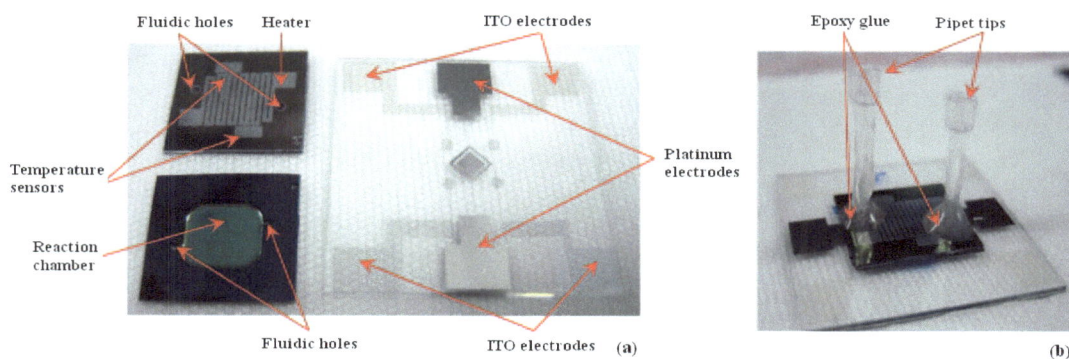

Figure 4: Photographs showing the silicon-glass microchip [8]. **(a)** Upper left: top view of the silicon chip showing the fluidic holes along with thin-film platinum heater and temperature sensors; lower left: bottom view of the silicon chip showing the 8 µL reaction chamber and the through-hole for fluid introduction; right panel: glass chip with patterned indium tin oxide working electrodes. **(b)** The assembled silicon-glass microchip with pipet tips glued to the fluidic holes. Reprinted with permission from the author. Copyright (2006).

This method of cell thermal lysis can be easily integrated into the PCR microchip. However, not all applications are compatible with thermal lysis. For example the method of thermal lysis is not suitable to protein analysis due to the denature of protein molecules at high temperatures.

2.4. Electrical-Based Lysis

Electroporation utilizes electric fields to create transient or permanent pore(s) on the cell membranes. This phenomenon occurs when sufficient voltage required for dielectric breakdown of the membrane, about 0.2~1.5 V, is imposed by an external electric field. Pores can be of two states: reversible electrical breakdown, with the pores resealing themselves when the external electric field is removed, and irreversible breakdown, where the damage is permanent.

Exposure of cells to strong electric fields can compromise cell membranes sufficiently to induce cell lysis. As cells are exposed to high-intensity pulsed electric fields, membranes are destabilized and become transiently permeable to macromolecules. As pulse length and electric field strength reach a critical value, the cell is lysed by dielectric breakdown of the cell membrane [34].

Various different bacteria and mammalian cells have different cell shapes and radius, which further lead to the variations in the critical electric field value for lysis [35]. In each case the applied electric field induced lysis when the cell membrane potential reached on average 1.1 V, usually requiring an external applied electric field of about 20 kV/cm [36]. Pulse length and number must also be set within critical parameters.

Lee *et al.* [11] reported an electrical method for cell lysis in microfluidic chips as an alternative to mechanical lysis, which was implemented based on the effect of electroporation. In this design, a micromachined cell lysis device with multi-electrode pairs to apply electric fields to cells was used. The schematic of the device is shown in (Fig. **5a**). Yeast cells were used to test this microdevice. First, the cells and the medium were pumped into the channel. Next, the cells were attracted to the sharp point of the electrode by dielectrophoretic forces using AC voltages in the frequency range of a few hundred kHz to a few MHz. Then, they were lyzed by pulsed electric fields. The edge of the electrodes was sharp in order to have a field concentration with the gaps of 5 µm. Parylene was used to make blocks between electrode pairs.

(Fig. **5b**) shows yeast cells that were not exposed to the electric fields and (Fig. **5c**) shows the cells after lysing. In (Fig. **5c**), only membrane debris was found after pulsed electric field treatment. The black dots in (Fig. **5c**) are cell debris. The optimum value for yeast cell lysing was achieved at 100 ms and 20 V. Then using this microdevice, cell lysis was achieved for Chinese cabbage, radish cells and *Escherichia coli* when a DC square pulse of only 3.5~20 V was used.

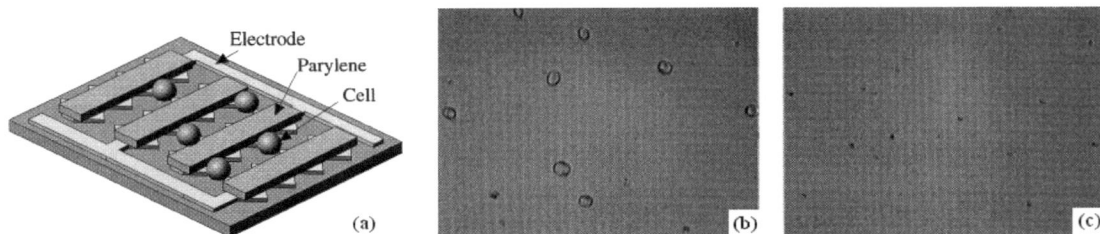

Figure 5: Schematic view of cell lysis device [11]. **(a)** and photograph of yeast cells before **(b)** and after **(c)** lysing. Reprinted with permission from Elsevier. Copyright (1999).

A similar design was demonstrated by Jensen *et al.* [37], where human carcinoma cells were disrupted electrically to release the subcellular materials. 74% cells were lyzed in the microdevices by using 8.5 V at 10 kHz AC current. The micro device was effective in lysing cells while operating at more advantageous conditions than conventional systems: small voltages and power consumptions, continuous flow, small sample volume, and negligible heating.

Later the same group further improved the electrode geometry to increase lysis efficiency [39]. Three-dimensional cylinder electrodes were designed and fabricated in an electroporation chip for cell lysis. This chip provides more volume in which the electric field affects the cell membrane.

Checkerboard or square-wall patterns of platinum electrodes were designed and fabricated on a silicon chip to separate *E. coli* from blood by dielectrophoresis before lysis [40]. Dielectrophoresis was accomplished by using a 10 V, 10 kHz AC current, while lysis was performed by a series of 400~500 V, 50 ms pulses with polarity alternated every 20 pulses. This lysis method allowed downstream analysis of RNA, plasmid DNA and genomic DNA by a hybridization assay, showing great potential in the area of micro total analysis system.

By carefully controlling the strength of the electrical field, microfabricated electroporation devices can also reversibly destabilize the cell membranes for gene transfection applications [41].

The electrode chip for cell lysis has the disadvantages such as low fabrication repeatability, short lifetime of the electrodes, which limited further application of this technique in both cell lysis and electroporation.

Recently many electric field type microfluidic chips were developed for cell disruption, coupled with Capillary Electrophoretic (CE) separation [12-13].

The group of Ramsey [38] proposed a microfluidic system, shown in (Fig. **6**), where, the voltages were applied to lyse the cells and also used to perform the electrophoretic injection for lysate separation. For the given design, both the steps of cell lysis and Capillary Electrophoretic (CE) separation were achieved on one single chip, and disruption duration of cells was as low as 33~40 ms. The integration of cell lysis chips with CE chips has great potential in medical diagnoses by comparing the electrophoretic peaks of the cellular debris of normal cells and that of abnormal cells.

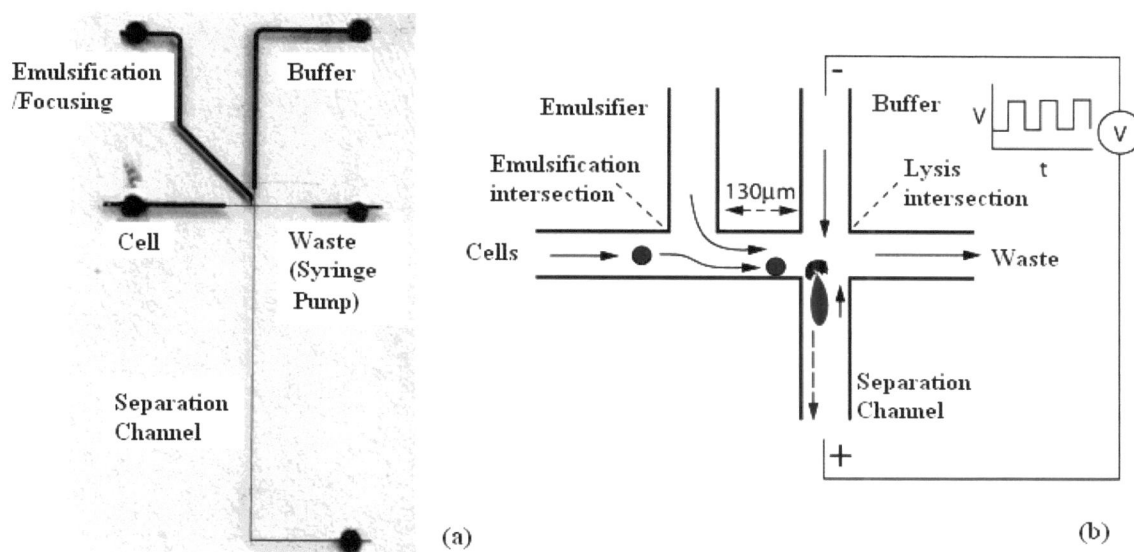

Figure 6: (a) Image of microchip used for the cell analysis experiments. **(b)** Schematic of the emulsification and lysis intersections for the microchip design shown in (a). The solid arrows show the direction of bulk fluid flow and the dashed arrow shows the electrophoretic migration direction of the labeled components in the cell lysate [38]. Reprinted with permission from the American Chemical Society. Copyright (2003).

Another study established an electric field by inserting platinum wires in PDMS microfluidic reservoirs connected by a channel which narrowed an order of magnitude in width to intensify the electric field and lyse the cells [42]. Local field strength of 1000-1500 V/cm was required for nearly 100% cell death.

Microfabricated system integrated with parallel electrical lysis and single-cell capillary electrophoresis was also reported by Munce [43]. Multiple parallel microchannels were designed and used to both increase throughput and eliminate cross-contamination between different separations.

Another method of electrical lysis involves using DC voltages to electrochemically lyse cells by generating hydroxide which cleaves the fatty acid groups and permanently breaks open the cells [44, 45].

In summary, the electrical lysis method is increasingly used in microfluidic devices, as a reagent-free, faster and less expensive alternative to chemical treatment which can be easily integrated with detection techniques including capillary electrophoresis and electrochemical detection.

2.5. Chemical Lysis

In chemical lysis, chemical/biological lysis buffers with reagents, such as SDS, TritonX-100, protein enzyme K, lysozyme are used to break down the cell walls and/or membranes [46-52]. The chemical cell lysis method, especially chemical dissolve method, has been used in the area of microfluidics due to easy miniaturization and integration with with other microfluidic components.

Li and Harrison [14] reported cell lysis on microdevices using Sodium Dodecyl Sulfate (SDS) based chemical method, along with the movement of cells in glass chips using electrokinetic transport. Although there are various lysing reagents (acidic solution, low concentration of copper (II) ions, deionized water) for red blood cells, the anionic detergent, SDS is capable of not only disrupting red blood cells sufficiently rapidly, but also modifying the electroosmotic flow rate. In this reported device, canine erythrocytes were disrupted after mixing a stream of cells at the double-T region with another stream containing a lysing agent of SDS by electrokinetic transport, shown in (Fig. 7). (Fig. 7a) shows cells entering from the left, with SDS mixing in from above. In (Fig. 7b), the two cells marked 1 and 4 begin to react with SDS and lyse. Since the SDS has not diffused across the whole channel, cells 2 and 3 along the wall are still intact by visual inspection. (Fig. 7c) shows that the remains of cell 1 are transported out of view, while the other three marked cells are also now lysed. A total time of 0.3 sec elapsed for these three frames.

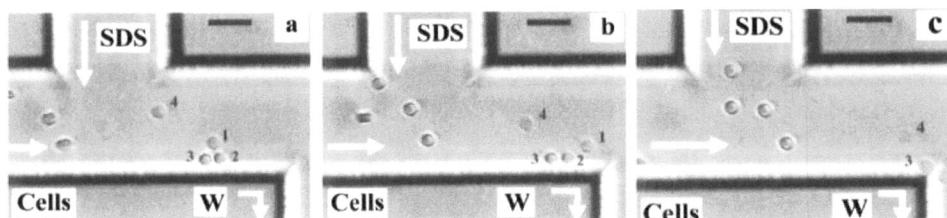

Figure 7: Photomicrographs of erythrocyte cell lysis in a microfluidic device [14]. White arrows show direction of flow and black bar shows the scale (20 μm). Cells enter from the left and SDS from above. Four cells are marked for reference in the text. A time progression over 0.3 s is illustrated in the three frames. Reprinted with permission from the American Chemical Society. Copyright (1997).

SDS is an ionic surfactant, which can quickly denature proteins, and is primarily used in DNA assays. Unfortunately, the use of SDS mightn't be comparable with enzymatic, immuno, or affinity based assays since the SDS binds to the proteins.

Schilling *et al.* [15] reported a relatively simple T-type microfluidic cell lysis chip based on chemical lysis. This device illustrated the ability of chemical lysis techniques to operate in a continuous flow mode. A microfluidic device with three inlets, two outlets and two central channels was designed and fabricated for the continuous lysis of bacterial cells and the fractionation/detection of intracellular proteins, shown in (Fig. 8). A cell suspension and a chemical lytic agent enter through separate inlets into the lysis channel. These two fluid streams flow side by side in the channel with no mixing except by lateral diffusion.

The continuous lysis of bacterial cells (*E. coli.*) was realized by using a proprietary mild detergent, β-galactosidase that did not affect the proteins with complete lysis of cells in less than 1 sec. In this experiment, a sample containing the cells was injected into a fluid channel where it met with another solution containing a chemical lysing agent. After the two fluids joined, they flowed down the lysis channel where the lysing agent diffused into the cell sample and disrupted the cell membranes. The intracellular components (some proteins of interest) left the disrupted cells and diffused out. These intracellular components were then free to diffuse in all directions and some diffused into the right half of the lysis channel. The fluid then came to a T-junction. All the cells including the intact cells and the permeated cells flowed out at the controlled outlet. At the same time, the lytic agent, along with some of the intracellular components, went in the other direction towards the detection channel. In the detection channel, a fluorescent product was produced through the reaction of a fluorogenic substrate with an intracellular enzyme such as β-galactosidase by their inter-diffuse in the detection channel.

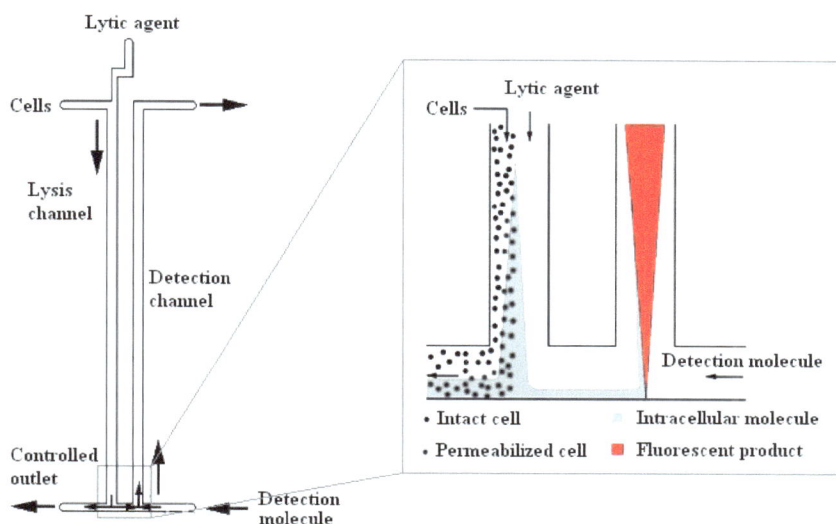

Figure 8: Schematic of microfluidic device for cell lysis and fractionation/detection of intracellular components. Pump rates are controlled at all inlets and one outlet. Lytic agent diffuses into the cell suspension, lysing the cells. Intracellular components then diffuse away from the cell stream and some are brought around the corner into the detection channel, where their presence can be detected by the production of a fluorescent species from a fluorogenic substrate [15]. Reprinted with permission from the American Chemical Society. Copyright (2002).

Finally the concentration of the particular intracellular component in the cells was detected in this continuous-flow, pressure-driven microfluidic device. For the given chip, fluorescent detection of the enzymatic reaction showed that cell lysis and protein separation could be sequentially implemented on a single chip.

Heo *et al.* [53] reported that SDS was used to lyze the *E. coli.* cells, which were immobilized in the PDMS chip first. *E. coli.* cells were trapped in a hydrogel in a fluid channel and then lyzed by passing 1% (w/v) SDS solution through the channel for 20 min. The trapped bacteria cells were successfully disrupted.

Another microfluidic device was developed for continuous cell lysing by using a chemical method reported by the group of Toner [54]. The microfluidic device with three inlets (blood inlet, lysis buffer inlet and PBS inlet), one outlet and a main serpentine lysis channel was designed and fabricated by typical soft lithography techniques. A lysis buffer containing mainly ammonium chloride was used to lyse the cells, which was introduced at the lysis buffer inlet. The single inlet of this lysis buffer was split into two channels that then met to sandwich the blood sample before the main serpentine lysis channel. The blood sample was pumped at the blood inlet and focused into a narrow stream flanked on both sides by the lysis buffer. The surface area of contact between the lysis buffer and the sample cells were increased to improve mixture efficiency and to reduce the lysis time. Red blood cells were disrupted after the blood sample and the lysis buffer was fully mixed during the flowing process in the long serpentine channel.

Finally, Phosphate-Buffered Saline (PBS) solution was pumped at the PBS inlet for restoration of normal physiological conditions. The single inlet of the PBS solution was again split into two channels that then met to sandwich the mixture of blood sample and the lysis buffer after the main serpentine lysis channel, which lead to the recovery of leukocytes.

In this experiment, complete lysis of erythrocytes and ~100% recovery of leukocytes were achieved, after the cells were exposed to an isotonic lysis buffer for less than 40 sec.

The mixture efficiency is a key factor for chemical cell lysis. A magnetically actuated micromixer was designed into a microfluidic device to disrupt cells by using the chemical lysis method [55]. More than 90% bacterial cells were successfully disrupted after starting the micromixer for 25 min.

In addition, a microdevice capable of cell lysis and DNA extraction from sperms was presented by Bienvenue [56] where a variety of chemical lysing agents were assessed in the extraction protocol.

The large number of microfluidic systems that use chemical means to lyse cells indicates the versatility of this method, especially to the application of cell lysis and DNA purification. However, the use of lysis reagents requires additional treatment to remove them from the cell lysates. Another main drawback is that this technical cannot be used for cell lysis and protection detection due to the denaturation of protein macromolecules caused by detergents.

2.6. Optical Cell Lysis

Recently, optical techniques were also applied for microfluidics based cell lysis. Cell membranes could be locally destroyed by using laser beams with a special wavelength.

For example, a novel cell lysis system was developed based on laser-induced disruption of bacterial and yeast cells [57]. First, *E. coli* cells as the model were used to get the optimal conditions including optical laser wavelength, lowest energy input. An average power of 100 mW for the lasers was shown to be sufficient to obtain cell lysis at wavelengths above 1000 nm, with optimal wavelengths between 1250 nm and 1550 nm. Under these conditions, the *E. coli* cells were broken down with RNA release. Using the parameters collected from bacteria lysis, *M. luteus*, *B. subtilis*, *B. cereus*, and *S. cerevisiae* were successfully disrupted by the laser-induced lysis technique.

Then potorous rat kidney epithelial cells were brokedown by a pulsed laser microbeam irradiation at $\lambda = 532$ nm in cell monolayers cultured at densities of 600 and 1000 cells/mm^2 with pulse energies of 5.6-24 μJ [59], which further demonstrates that the optical method of laser-induced cell lysis is a reliable method that can be used for sample preparation, without the concern of damage to macromolecules.

Another optical lysis microfluidic platform was developed by using the 808 nm laser and nanoparticles (gold nanorods) [60]. The thermal energy being able to disrupt cells was transformed from the light source by using Au nanorods in a microfluidic chip. The laser absorbance capability of gold nanorods is different with their different shapes. The optimal rod shape (aspect ratio 3) was found to produce high efficient heat absorption for optimal lysis. Consequently the DNA was extracted out of the cells and transferred to a PCR system.

Then the same group reported a Compact Disk (CD) microfluidic platform by incorporating above laser-induced cell lysis system [58]. A fully integrated, pathogen-specific DNA extraction device with optical cell lysis was designed and fabricated by using centrifugal microfluidics on a polymer based CD platform. The design principles were shown schematically in (Fig. **9a**). Magnetic particles conjugated with target specific antibodies were mixed with sample solutions. Target pathogens were selectively captured on the magnetic beads and the waste materials such as plasma residue were washed away. The laser beam (808 nm, 1.5 W) was effectively absorbed on magnetic particles and the heat was generated very rapidly. Simple irradiation by the laser for 30 sec effectively disrupted cells and extracted PCR-ready DNA from captured target pathogens.

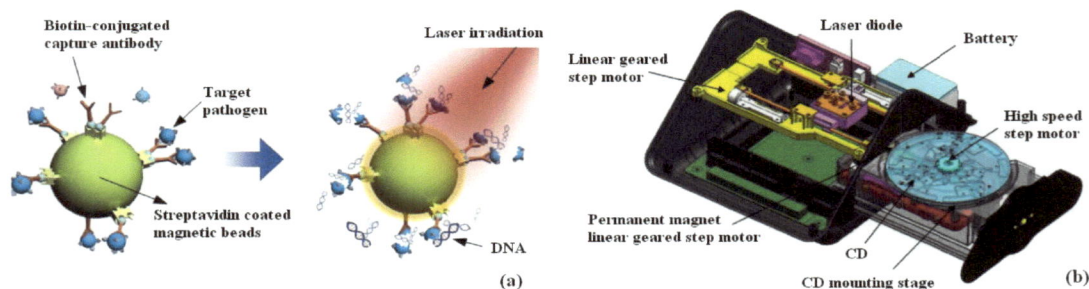

Figure 9: (a) Schematic diagram of the reaction principle [58]. **(b)** Schematic diagram showing the inside of the portable lab-on-a-disc device. A laser diode was mounted on a linear geared step motor. A permanent magnet was located on the other linear geared step motor located under the CD mounting stage. A high speed step motor was used to run the spin program. Reprinted with permission from the Royal Society of Chemistry. Copyright (2007).

A hand-held type sample preparation device was developed as shown in (Fig. **9b**). Using this device, cell lysis and nucleic acid sample preparation were successfully performed on a single device. First Hepatitis B Virus (HBV) and *E. coli* cells were captured from whole blood and then broken down under the laser irradiation. Finally the DNA molecules of Hepatitis B Virus (HBV) and *E. coli* cells were extracted on the single CD microfluidic device.

3. CASE STUDY: ON-CHIP CELL LYSIS *VIA* A CHEMICAL METHOD

In this case study, a continuous flow microfluidic biochip for blood cell lysis was designed, fabricated and characterized. On-chip cell lysis was conducted by the sufficient mixture of blood samples with lysing reagents [61-63].

In this example, the performance of two mixing models, namely T-type mixing model and sandwich-type mixing model, was compared with numerical simulations. Results of simulations indicated that the sandwich-type mixing model with coiled channels can perform better than the T-type counterpart, which lead to the further use of this model to construct the microfluidic biochip for continous cell lysis.

Rat blood with anticoagulant was the cell sample, while guanidine and Triton X-100 were used as the lysing reagents, respectively, with the effects of the two reagents on cell lysis efficiency compared. The effects of the cell concentration and the flow rate on cell lysis were analyzed using guanidine as the lysing reagent. Blood cells can be lysed in a few minutes when the flow rate of the lysing reagent was considerably higher than that of the cell sample. This for cell lysis microfluidic device has the potential to integrate with other microfluidic components for cell separation and DNA extraction, enabling an integrated microfluidic system for whole blood sample pretreatment.

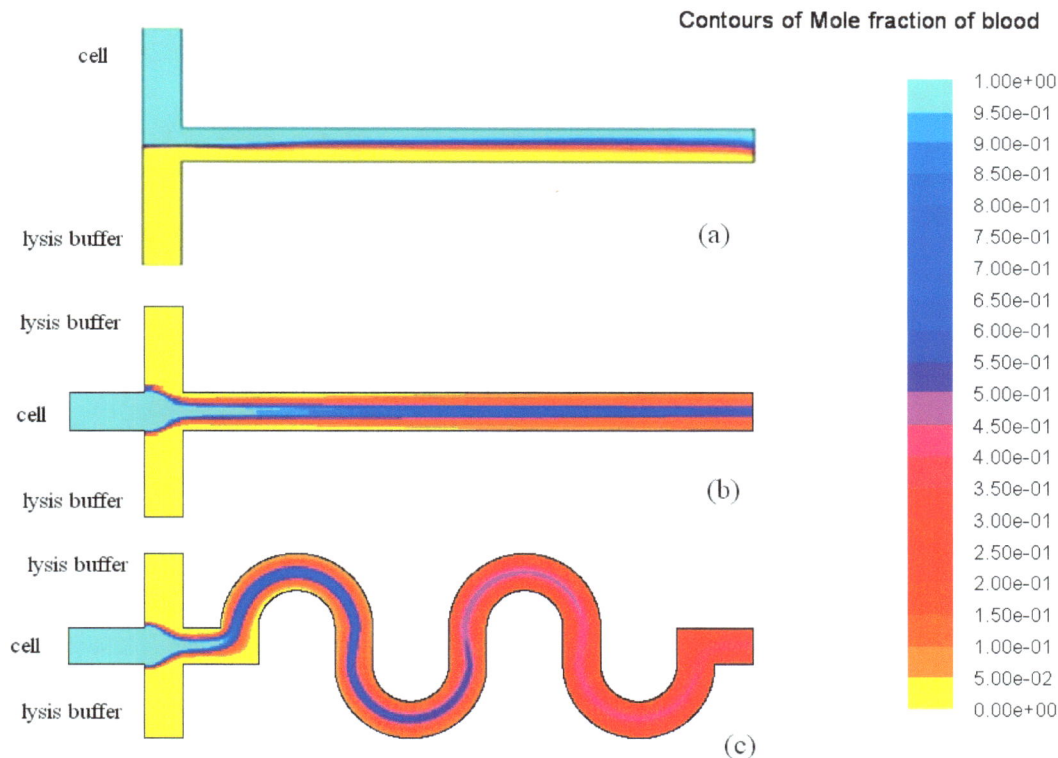

Figure 10: Numerical simulation results for the species concentration distributions at T-type mixing model (**a**), sandwich type mixing model with lined channel (**b**) and sandwich type mixing model with coiled channel (**c**) when v_{cell} = 0.005m/s, v_{buffer} = 0.0025m/s. The color for "1.00e+00" and "0.00 e+00" in this figure refers to the concentration intensity of the pure cell sample and the pure lysis buffer, respectively.

3.1. Method

Although the thermal, ultrasonic, and electrical approaches for cell lysis have been shown to work very well, they all demand the use of external power supplies, which renders devices complicated and costly to fabricate. In the meanwhile, chemical based cell lysis can be easy integrated with microfluidics by using sandwich fluid flows in microfluidic channels and therefore, it has been used a lot for cell lysis and DNA purification.

DNA of blood cells in the karyon is bonded with nucleoprotein, and it is not released under the normal conditions. To release DNA in the karyon, lysing reagents need to dissolve the cell membranes and karyotheca, making DNA detach from the denatured nucleoproteins. Guanidine is the bond reagent for extracting DNA by means of solid phase extraction, and is also a generic strong denaturant of the nucleoproteins. In this example, guanidine was used as the lysis reagent of blood cells, and the effect of the cell concentration and the flow rate on the cell lysis was studied.

(Fig. **10**) shows that the T-type mixing model and the sandwich type mixing model used in this study. In the sandwich-type model, the line-type structure, and the coil-type structure of the mixing microchannels were compared. All the mixing elements were fabricated from silicon substrates. The experimental conditions and model geometries were optimized by numerical simulations and verified by experiments.

Computational Fluid Dynamics (CFD) simulations were conducted before device fabrication and characterization since numerical simulation enables the relevant parameters to be varied systematically over a wide range. In general, numerical simulation plays a key role in device design optimization and can also be used to interpret experimental results.

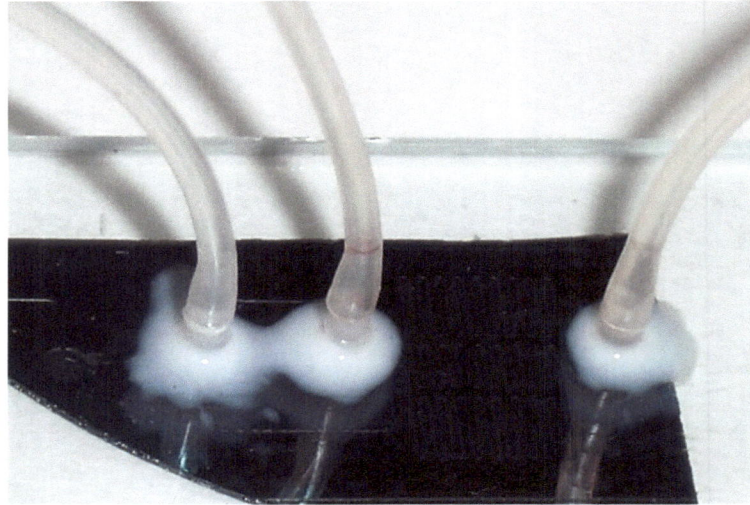

Figure 11: Photograph of the microfluidic biochip including a cell inlet, a buffer inlet and an outlet.

All simulations were two-dimensional using the laminar flow model. The numerical simulation results are shown in (Fig. **10**). The mixing performance in the sandwich-type design was better than that of the T-type counterpart because of the increase in the contact area. The mixing performance in the coiled channel was much better than that in the lined channel. These numerical simulation results mentioned above provide a very clear understanding of the physical phenomena taking place in the 2-D microfluidic channels. In order to quantify the mixing performance of these designs, a mixing index (s) is used as follows [27]:

$$\sigma = \left(1 - \frac{\int_0^h |c - c_\infty| \, dy}{\int_0^h |c_0 - c_\infty| \, dy} \right) \times 100\% \qquad (1)$$

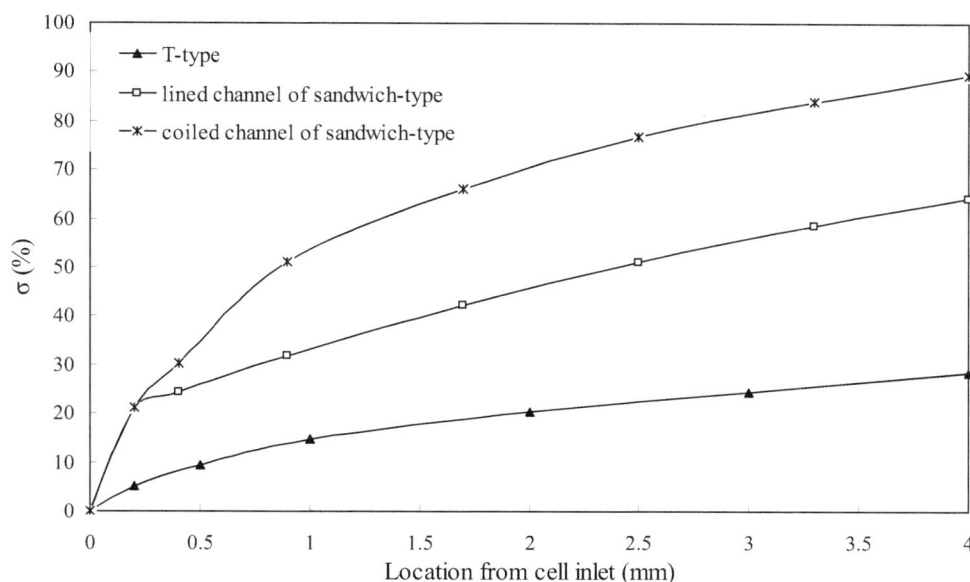

Figure 12: Numerical evaluations of mixing efficiency index (σ) at different mixing models when v_{cell} = 0.005m/s, v_{buffer} = 0.0025m/s.

where c is the species concentration profile across the width of the microchannel (h), c_∞ is the completely mixed state (= 0.5) and c_0 is the completely unmixed state (= 0 or 1). Note that the confluent streams are completely mixed if σ = 100%. In contrast, they are completely unmixed if σ = 0%. (Fig. **11**) shows the mixing efficiency of the proposed models: T-type models, sandwich-type models with lined channels, and sandwich-type models with coiled channels. The use of coiled channels effectively enhance chaotic mixing, with the mixing efficiency of 89.3% obtained at the cross section located 4 mm from the inlet, by folding, stretching, and reorienting fluid. Based on the discussion above, the coiled channel of the sandwich-type mixing model was chosen to fabricate microfluidic chip.

For chemical cell lysis, there are two steps in total, which are mixing and chemical reaction, respectively. In the microchip, the mixing process is slow compared with the fast chemical reaction stage due to the laminar flow in microfluidic channels. The total lysis of cells requires that each cell is fully exposed to the lysis buffer, which is limited by the mixing speed. Thus, the cell lysis efficiency strongly depends on the mixing performance between the cell solution and the lysis buffer. Since the sandwich-type mixing model enables good mixture of cells with the buffer solution demonstrated by numerical simulations, this design was fabricated for experiments.

As shown in (Fig. **12**), the microchip consists of a glass cover with two inlets and one outlet and a silicon substrate with an etched coiled channel of 200 μm in width, 100 μm in depth, and 20.16 cm in length. The cell solution was sandwiched between the cell lysis buffers. The microchip was mounted onto a stage of a microscope with a CCD camera and a video monitoring system. Blood was introduced through the cell inlet by a peristaltic pump, while the lysis buffer (namely, load buffer) was pumped through the buffer inlet by another peristaltic pump. When the lysis buffer and blood were mixtured in the channel during the continuous flow, cells were gradually lyzed, shown in (Fig. **13**).

3.2. Results and Discussion

3.2.1. Effect of the Lysing Reagent

Based on the different biological samples and the requirements of detection, different lysis reagents were used. Actually, there are a few lysis reagents for choose in the area of blood cell lysis, such as proteinase K, SDS and Triton X-100. In addition, based on the principle of the osmotic shock, water can be directly used for lyzing blood cells, since the solute concentration of blood cells is higher than the pure water. When water enters the cells, the cells are swelled and finally the cells are damaged. The solute concentration of cells is approximately 0.1 M or 0.2 M NaCl. Thus, the osmotic pressure is high enough to completely lyze cells. Proteinase K is used for cell lysis since the cell wall is digested by this enzyme. The advantage of this method is that there is no damage to the other substances in the cells, under gentle conditions. However, the disadvantage is that the enzyme is expensive. SDS and Triton X-100 are generic surfactants, having amphoteric chemical properties, which react with water and can also react with lipid. These generic lysis reagents are used to lyze cells by dissolving lipid in the cell wall exposed to surfactants [52, 64].

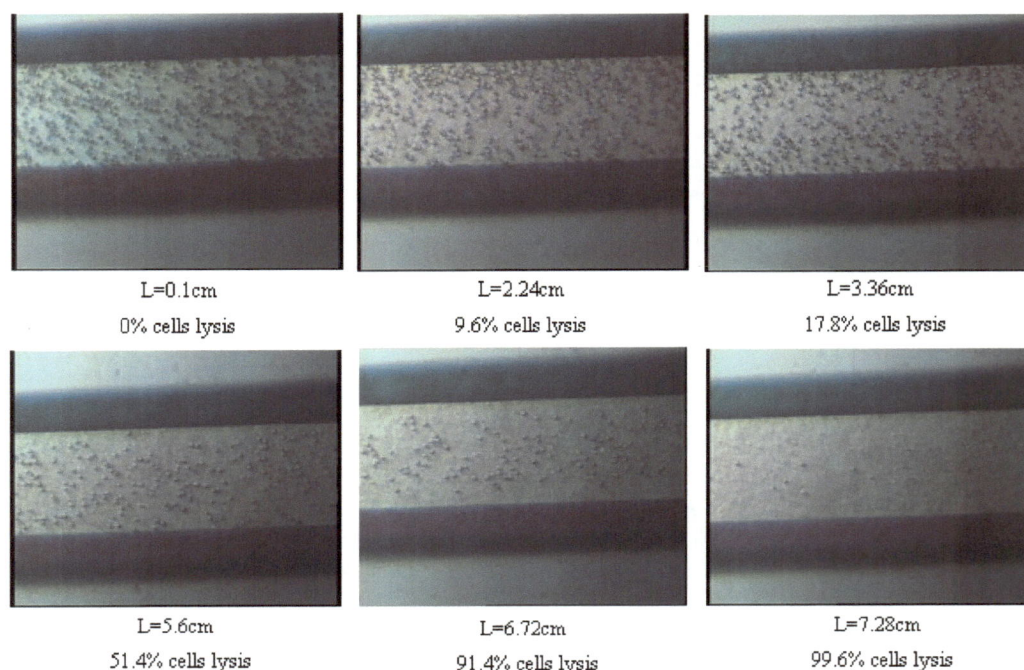

L=0.1cm	L=2.24cm	L=3.36cm
0% cells lysis	9.6% cells lysis	17.8% cells lysis
L=5.6cm	L=6.72cm	L=7.28cm
51.4% cells lysis	91.4% cells lysis	99.6% cells lysis

Figure 13: Photograph of cell lyzing in microfluidic chip L: the distance traveled in the microchannels, %: (the percentage of lysed cells).

High concentration guanidine salt is generally used in porous solid phase matrix for the extraction of the nucleic acids and is also a generic strong denaturant. Guanidine salt is capable of quickly dissolving protein and causing damage to cell structures. By using guandinie salt, nucleoprotein is also unbound from nucleic acid because of the damaged secondary structure. The high salt concentration also allows the salt bridge that forms to extract RNA in silica environments [65, 66].

In this example, guanidinium thiocyanate (lysis buffer A) was used to lyze the rat peripheral blood samples that were diluted 100-fold compared to raw samples. The flow rates of the lysis buffer and the cell sample were both 0.2 μl/min. Blood cells were introduced into the microchip flanked by the lysis buffer on both sides and continuously flowing in the microchannel. During the continuous flow, blood cells were lyzed gradually, as shown in (Fig. **13**). Experimental results show that after blood cells flowed past the specific checking point, which was 7.28 cm from the inlet along the channel, 99.6% blood cells were lyzed. Based on the dimension of the microchannel, it was calculated that the time taken for complete blood cell lysis was approximately 3 min. In addition, three lysis buffers, namely, Guanidine salt (lysis buffer A), TritonX-100 (lysis buffer B) and the mixture (lysis buffer C) were used for the cell lysis experiment under the same experimental conditions for comparison. Guanidine salt (lysis buffer A) enables blood cell lysis and DNA unbound from nucleaproteins, which can be used for DNA purification directly. The results are shown in (Fig. **14a**). Based on the dimensions of the microchannel, it was calculated that the time taken for complete blood cell lysis was 174 sec by guanidine salt, 161 sec by TritonX-100, 108 sec by the mixture buffer C. According to the experimental results, cell lysis by the lysis buffer B was almost comparable to the lysis buffer A while the performance of cell lysis using the mixture buffer is slightly higher than individual cell lysis buffers.

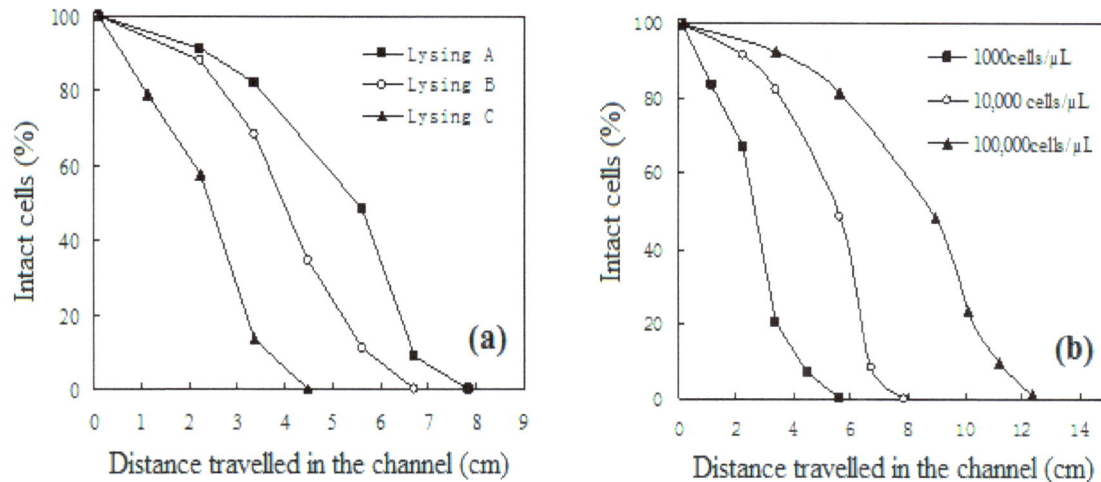

Figure 14: **(a)** Effect of different lysing reagents on cells lysing. **(b)** Effect of cell concentration on cell lysing.

3.2.2. Effect of the Cell Concentration

Cell lysis experiments were implemented with the rat peripheral blood samples, which were diluted 10, 100, and 1000 times, respectively, with the lysis reagent of guanidine salt (lysis buffer A). The flow rates of the lysis buffer and the cell sample were both 0.2 μl/min. The results are shown in (Fig. **14b**). With the increase of the cell concentration, the process for cell lysis was prolonged. The rat blood sample diluted 1000-fold was completely lyzed at the checking point, which was 0.56 cm away from the inlet along the channel, while the rat blood sample diluted 10-fold was completely lyzed at the specific location, which was 1.34 cm from the inlet along the channel. The time taken for lyzing rat blood diluted 1000-fold was 134 sec while the time taken for lyzing rat blood diluted 10-fold was more than 323 sec by calculation.

3.2.3. Effect of the Cell Flow Rate

The effect of the cell flow rate on the cell lysis was conducted with the rat peripheral blood sample diluted 100 times and the lysis reagent of guanidine salt (lysis buffer A). The cell lysis experiments were conducted under the condition that the flow rate of the cell sample was the same as that of the lysis buffer. The results are shown in (Fig. **15a**). With the increase of the cell flow rate, the distance traveled by cells before full lysis in the microchannel increased. When the cell flow rate was higher than 0.5 μl/min, blood cells were not completely lyzed in the microchip (Fig. **15a**). The increased cell flow rate resulted in the decrease in the flow time of solutions in the microfluidic chip and the reduction of the chances of cells exposed to the lysis buffer, and therefore producing to a lower efficiency in cell lysis. However, experiments show that the time for cell lysis was insensitive to the cell flow rate and was always approximately 180 sec (Fig. **15c**). Under the condition of the lysis buffer of 1 μl/min, cell lysis experiments were implemented at different cell flow rates. The results are shown in (Fig. **15b** and **15d**). With the increase of the cell flow rate, the distance of cells traveled in the microchannel increased, which indicated that the time taken for cell lysis was also prolonged. From experiments, the performance of cell lysis was improved considerably when the lysis buffer flow rate is larger than the cell flow rate. When the ratio of the lysis buffer flow rate and the cell flow rate was 5~10, the time for complete cell lysis was only a few seconds (Fig. **15 d**). When the lysis buffer flow rate was higher than the cell flow rate, the chances of the cells exposed to the lysis buffer were higher, which made the cell lysis process shorter. Under the condition that the ratio of the lysis buffer flow rate and the cell flow rate is higher than 5, cell lysis can be fleetly accomplished in the microchip by improving the flow rate of both the blood cell sample and the lysis buffer.

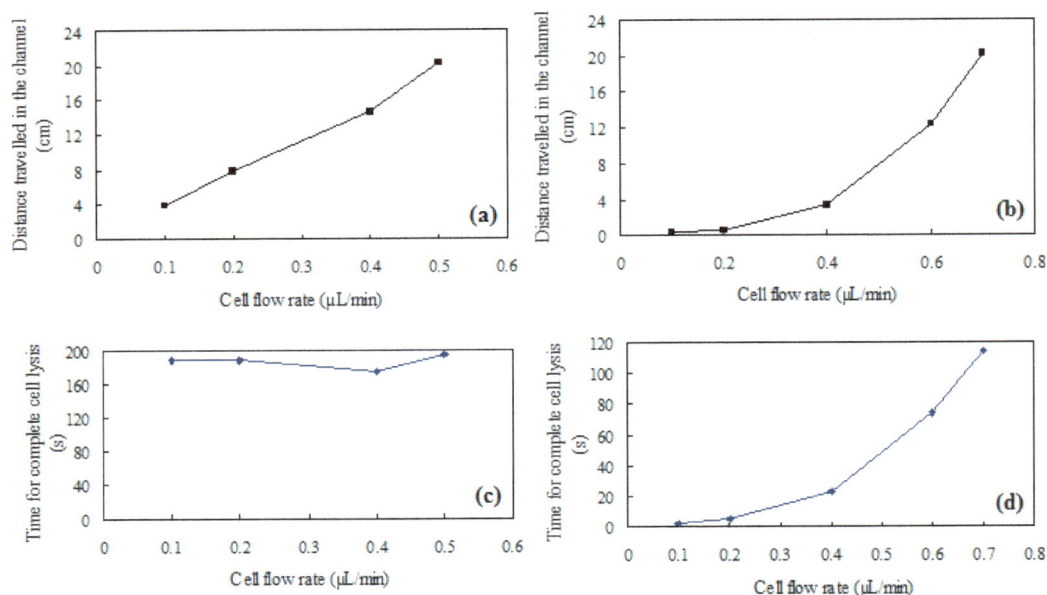

Figure 15: Relation between cells flow rate and distance traveled in the channel (**a, b**). Relation between cells flow rate and time for complete cell lysis (**c, d**).

4. CONCLUSION

Experiment results demonstrated that guanidine salt is capable of cell lysis due to its ability to denature protein, with comparable performance to traditional surfactants. By keeping the ratio of the lysis buffer flow rate and the cell flow rate higher than 5, rapid cell lysis was achieved. Since guanidine has been used for porous solid phase matrix based nucleic acid extraction, the sandwich flow cell lysis chip using guanidine as the lysis reagent can be easily integrated with other components such as the solid phase extraction chip for nucleic acid extraction, which will lay the foundation for the full realization of μTAS.

ACKNOWLEDGMENT

The authors greatly acknowledge the financial support from the National Science Foundation of China under Grant numbers 60701019 and 60501020.

REFERENCES

[1]   Chassy B, Giuffrida A. Method for the lysis of Gram-positive, asporogenous bacteria with lysozyme. Appl Environ Microbiol 1980; 39(1): 153-8.
[2]   Birnboim H, Doly J. A rapid alkaline extraction procedure for screening recombinant plasmid DNA. Nucleic Acids Res 1979; 7(6): 1513-23.
[3]   Kondo T, Gamson J, Mitchell J, Riesz P. Free radical formation and cell lysis induced by ultrasound in the presence of different rare gases. Int J Radiat Biol 1988; 54(6): 955-62.
[4]   Moré M, Herrick J, Silva M, Ghiorse W, Madsen E. Quantitative cell lysis of indigenous microorganisms and rapid extraction of microbial DNA from sediment. Appl Environ Microbiol 1994; 60(5): 1572-80.
[5]   Carlo D, Jeong K, Lee L. Reagentless mechanical cell lysis by nanoscale barbs in microchannels for sample preparation. Lab Chip 2003; 3(4): 287-91.
[6]   Taylor MT, Belgrader P, Furman BJ, *et al.* Lysing Bacterial spores by sonication through a flexible interface in a microfluidic system. Anal Chem 2001; 73(3): 492-6.
[7]   Belgrader P, Hansford D, Kovacs G, *et al.* A minisonicator to rapidly disrupt bacterial spores for DNA analysis. Anal Chem 1999; 71(19): 4232-6.

[8]   Yeung SW, Lee TMH, Cai H, Hsing IM. A DNA biochip for on-the-spot multiplexed pathogen identification. Nucleic Acids Res 2006; 34(18): e118.

[9]   He Y, Zhang YH, Yeung ES. Capillary-based fully integrated and automated system for nanoliter polymerase chain reaction analysis directly from cheek cells. J Chromatogr A 2001; 924(1-2): 271-84.

[10] Waters LC, Jacobson SC, Kroutchinina N, *et al.* Microchip device for cell lysis, multiplex PCR amplification, and electrophoretic sizing. Anal Chem 1998; 70(1): 158-62.

[11] Lee SW, Tai YC. Micro cell lysis device. Sensor Actuator: A: Phys 1999; 73(1-2): 74-9.

[12] Gao J, Yin XF, Fang ZL. Integration of single cell injection, cell lysis, separation and detection of intracellular constituents on a microfluidic chip. Lab Chip 2004; 4(1): 47-52.

[13] Han F, Wang Y, Sims C, *et al.* Fast electrical lysis of cells for capillary electrophoresis. Anal Chem 2003; 75(15): 3688-96.

[14] Li PCH, Harrison DJ. Transport, manipulation, and reaction of biological cells on-chip using electrokinetic effects. Anal Chem 1997; 69(8): 1564-6.

[15] Schilling EA, Kamholz AE, Yager P. Cell lysis and protein extraction in a microfluidic device with detection by a fluorogenic enzyme assay. Anal Chem 2002; 74(8): 1798-804.

[16] Takamatsu H, Takeya R, Naito S, Sumimoto H. On the mechanism of cell lysis by deformation. J Biomech 2005; 38(1): 117-24.

[17] Kim Y, Kang J, Park S, Yoon E, Park J. Microfluidic biomechanical device for compressive cell stimulation and lysis. Sensor Actuator B Chem 2007; 128(1): 108-16.

[18] Shen A. Granular fingering patterns in horizontal rotating cylinders. Phys Fluids 2002; 14(2): 462-70.

[19] Jalali P, Li M, Ritvanen J, Sarkomaa P. Intermittency of energy in rapid granular shear flows. Chaos 2003; 13(2): 434-43.

[20] Kim J, Jang SH, Jia G, *et al.* Cell lysis on a microfluidic CD (compact disc). Lab Chip 2004; 4(5): 516-22.

[21] Knoll S, Mulfinger S, Niessen L, Vogel R. Rapid preparation of Fusarium DNA from cereals for diagnostic PCR u sing sonification and an extraction kit. Plant Pathol 2002; 51(6): 728-34.

[22] Doebler RW, Erwin B, Hickerson A, *et al.* Continuous-flow, rapid lysis devices for biodefense nucleic acid diagnostic systems. JALA 2009; 14(3): 119-25.

[23] Marentis T, Kusler B, Yaralioglu G, *et al.* Microfluidic sonicator for real-time disruption of eukaryotic cells and bacterial spores for DNA analysis. Ultrasound Med Biol 2005; 31(9): 1265-77.

[24] Zhang P, Zhang G, Wang W. Ultrasonic treatment of biological sludge: Floc disintegration, cell lysis and inactivation. Bioresour Technol 2007; 98(1): 207-10.

[25] Sacks PG, Miller MW, Church CC. The exposure vessel as a factor in ultrasonically-induced mammalian cell lysis--I. A comparison of tube and chamber systems. Ultrasound Med Biol 1982; 8(3): 289-93, 95-98.

[26] Church CC, Miller MW. The kinetics and mechanics of ultrasonically-induced cell lysis produced by non-trapped bubbles in a rotating culture tube. Ultrasound Med Biol 1983; 9(4): 385-93.

[27] Feril LB, Kondo T, Zhao Q-L, *et al.* Enhancement of ultrasound-induced apoptosis and cell lysis by echo-contrast agents. Ultrasound Med Biol 2003; 29(2): 331-7.

[28] Brayman AA, Doida Y, Miller MW. Apparent contribution of respiratory gas exchange to the *in vitro* "cell density effect" in ultrasonic cell lysis. Ultrasound Med Biol 1992; 18(8): 701-14.

[29] Ogino C, Farshbaf Dadjour M, Takaki K, Shimizu N. Enhancement of sonocatalytic cell lysis of Escherichia coli in the presence of TiO2. Biochem Eng J 2006; 32(2): 100-5.

[30] Khanna P, Ramachandran N, Yang J, *et al.* Nanocrystalline diamond microspikes increase the efficiency of ultrasonic cell lysis in a microfluidic lab-on-a-chip. Diamond Relat Mater 2009; 18(4): 606-10.

[31] Brayman AA, Miller MW. Bubble cycling and standing waves in ultrasonic cell lysis. Ultrasound Med Biol 1992; 18(4): 411-20.

[32] Liu RH, Yang J, Lenigk R, Bonanno J, Grodzinski P. Self-Contained, fully integrated biochip for sample preparation, polymerase chain reaction amplification, and DNA microarray detection. Anal Chem 2004; 76(7): 1824-31.

[33] Lee CY, Lee GB, Lin JL, Huang FC, Liao CS. Integrated microfluidic systems for cell lysis, mixing/pumping and DNA amplification. J Micromech Microeng 2005; 15(6): 1215-23.

[34] Hamilton W, Sale A. Effects of high electric fields on microorganisms. II. Mechanism of action of the lethal effect. Biochim Biophys Acta 1967; 148(3): 789-800.

[35] Hülsheger H, Potel J, Niemann E. Electric field effects on bacteria and yeast cells. Radi Environ Biophys 1983; 22(2): 149-62.

[36]   Grahl T, Märkl H. Killing of microorganisms by pulsed electric fields. Appl Microbiol Biotechnol 1996; 45(1): 148-57.

[37]   Lu H, Schmidt MA, Jensen KF. A microfluidic electroporation device for cell lysis. Lab Chip 2005; 5(1): 23-9.

[38]   McClain MA, Culbertson CT, Jacobson SC, *et al.* Microfluidic devices for the high-throughput chemical analysis of cells. Anal Chem 2003; 75(21): 5646-55.

[39]   Lu KY, Wo AM, Lo YJ, *et al.* Three dimensional electrode array for cell lysis *via* electroporation. Biosens Bioelectron 2006; 22(4 SPEC. ISS.): 568-74.

[40]   Cheng J, Sheldon EL, Uribe A, *et al.* Preparation and hybridization analysis of DNA/RNA from *E. coli* on microfabricated bioelectronic chips. Nat Biotechnol 1998; 16(6): 541-6.

[41]   Lin Y, Jen C, Huang M, Wu C, Lin X. Electroporation microchips for continuous gene transfection. Sensors Actuator B Chem 2001; 79(2-3): 137-43.

[42]   Wang HY, Bhunia AK, Lu C. A microfluidic flow-through device for high throughput electrical lysis of bacterial cells based on continuous dc voltage. Biosens Bioelectron 2006; 22(5): 582-8.

[43]   Munce NR, Li J, Herman PR, Lilge L. Microfabricated system for parallel single-cell capillary electrophoresis. Anal Chem 2004; 76(17): 4983-9.

[44]   Carlo D, Ionescu-Zanetti C, Zhang Y, Hung P, Lee L. On-chip cell lysis by local hydroxide generation. Lab Chip 2005; 5(2): 171-8.

[45]   Nevill J, Cooper R, Dueck M, Breslauer D, Lee L. Integrated microfluidic cell culture and lysis on a chip. Lab Chip 2007; 7(12): 1689-95.

[46]   Chien LJ, Lee CK. Synergistic effect of co-expressing D-amino acid oxidase with T7 lysozyme on self-disruption of Escherichia coli cell. Biochem Eng J 2006; 28(1): 17-22.

[47]   Hall JA, Felnagle E, Fries M, *et al.* Evaluation of cell lysis procedures and use of a micro fluidic system for an automated DNA-based cell identification in interplanetary missions. Planet Space Sci 2006; 54(15): 1600-11.

[48]   Berezovski MV, Mak TW, Krylov SN. Cell lysis inside the capillary facilitated by transverse diffusion of laminar flow profiles (TDLFP). Anal Bioanal Chem 2007; 387(1): 91-6.

[49]   Stachowiak JC, Shugard EE, Mosier BP, *et al.* Autonomous microfluidic sample preparation system for protein profile-based detection of aerosolized bacterial cells and spores. Anal Chem 2007; 79(15): 5763-70.

[50]   El-Ali J, Gaudet S, Gunther A, Sorger PK, Jensen KF. Cell stimulus and lysis in a microfluidic device with segmented gas−liquid flow. Anal Chem 2005; 77(11): 3629-36.

[51]   Marc PJ, Sims CE, Bachman M, Li GP, Allbritton NL. Fast-lysis cell traps for chemical cytometry. Lab Chip 2008;8(5):710-6.

[52]   Pang Z, Al-Mahrouki A, Berezovski M, Krylov SN. Selection of surfactants for cell lysis in chemical cytometry to study protein-DNA interactions. Electrophoresis 2006; 27(8): 1489-94.

[53]   Heo J, Joseph Thomas K, Hun Seong G, Crooks RM. A microfluidic bioreactor based on hydrogel-entrapped E. coli: Cell viability, lysis, and intracellular intracellular enzyme reactions. Anal Chem 2003; 75(1): 22-6.

[54]   Sethu P, Anahtar M, Moldawer LL, Tompkins RG, Toner M. Continuous flow microfluidic device for rapid erythrocyte lysis. Anal Chem 2004; 76(21): 6247-53.

[55]   Huh YS, Choi JH, Park TJ, *et al.* Microfluidic cell disruption system employing a magnetically actuated diaphragm. Electrophoresis 2007; 28(24): 4748-57.

[56]   Bienvenue JM, Duncalf N, Marchiarullo D, Ferrance JP, Landers JP. Microchip-based cell lysis and DNA extraction from sperm cells for application to forensic analysis. J Foren Sci 2006; 51(2): 266-73.

[57]   Dhawan MD, Wise F, Baeumner AJ. Development of a laser-induced cell lysis system. Anal Bioanal Chem 2002; 374(3): 421-6.

[58]   Cho YK, Lee JG, Park JM, *et al.* One-step pathogen specific DNA extraction from whole blood on a centrifugal microfluidic device. Lab Chip 2007; 7(5): 565-73.

[59]   Rau K, Quinto-Su P, Hellman A, Venugopalan V. Pulsed laser microbeam-induced cell lysis: time-resolved imaging and analysis of hydrodynamic effects. Biophys J 2006; 91(1): 317-29.

[60]   Cheong KH, Yi DK, Lee JG, *et al.* Gold nanoparticles for one step DNA extraction and real-time PCR of pathogens in a single chamber. Lab Chip 2008; 8(5): 810-3.

[61]   Chen X, Cui D, Liu C, Cai H. Microfluidic biochip for blood cell lysis. Chin J Anall Chem 2006; 34(11): 1656-60.

[62]   Chen X, Cui D, Liu C, Li H, Chen J. Continuous flow microfluidic device for cell separation, cell lysis and DNA purification. Anal Chim Acta 2007; 584(2): 237-43.

[63] Chen X, Cui DF, Liu CC. On-line cell lysis and DNA extraction on a microfluidic biochip fabricated by microelectromechanical system technology. Electrophoresis 2008; 29(9): 1844-51.

[64] Bontoux N, Dauphinot L, Vitalis T, *et al.* Integrating whole transcriptome assays on a lab-on-a-chip for single cell gene profiling. Lab Chip 2008 ; 8(3): 443-50.

[65] Boom R, Sol C, Salimans M, *et al.* Rapid and simple method for purification of nucleic acids. J Clin Microbiol 1990; 28(3): 495-503.

[66] von Hippel P, Wong K. Neutral salts: The generality of their effects on the stability of macromolecular conformations. Science 1964; 145(3632): 577-80.

CHAPTER 5

Microfluidic Chips for DNA Extraction and Purification

Xing Chen[*], Dafu Cui and Jian Chen

State Key Laboratory of Transducer Technology, Institute of Electronics, Chinese Academy of Sciences, Beijing, China

Abstract: Nucleic acid extraction is an important element of sample preparation for genomic analysis which requires extracting DNA or RNA and removing other proteins or impurities. In this chapter, we firstly review traditional large-scale methods for DNA extraction and purification, followed by corresponding microfluidics based approaches. And then an example of DNA extraction from whole blood and culture cells in a microfluidic chip is provided in which the device design, fabrication and testing, based on the principle of solid phase extract (SPE) are covered in detail. In this specific example, various solid phase matrixes have been prepared, characterized and tested.

Keywords: DNA extraction, solid phase extraction, porous matrix, MEMS, microfluidics.

1. INTRODUCTION

For most of genetic analytical processes, the DNA must be amenable to amplification at a reasonably high concentration without endogenous PCR inhibitors. Purification of intact DNA/RNA is the primary step of many molecular biology techniques, including northern blotting, quantitative polymerase chain reaction, and microarray assays. DNA/RNA extraction is typically conducted using either a phenol-chloroform or a solid phase method. However, the traditional phenol extraction or solid phase extraction is a complex and time-consuming process because of the complex of crude biological samples utilizing manual processes with lab-scale equipment such as shakers and centrifuges, even by using some commercial purification kits. In addition, researchers may be under great risk of contamination from using organic reagents (*e.g.*, phenol–chloroform reagent) in the extraction processes and the bio-samples may be wasted during these operations.

Microfluidic methods have several advantages over their large-scale counterparts, including lower cost, disposability, portability, lower reagent and sample consumption, automation, and lower power consumption. Miniature biomedical devices for extraction and purification have also been widely explored by a number of researchers in the area of microfluidics.

In this chapter, we first review the traditional methods for DNA/RNA purification, and then focus on the reported miniaturized methods for DNA/RNA extraction. Finally we give a specific example to illuminate how to design, fabricate and characterize microfluidic devices for DNA extraction from whole blood samples and cell lines using Solid Phase Extraction (SPE). In this example, porous silicon as the solid phase matrix for DNA absorption was fabricated in microfluidic chips using MEMS technologies and the anodization technique. The fabrication process of porous silicon was investigated and optimized and the fabricated porous structures were characterized and evaluated in detail.

2. TRADITIONAL METHODS FOR DNA EXTRACTION

The most popular traditional method of nucleic acid extraction is the protocol with a phenol/chloroform alcohol solution. After cells are disrupted by chemical reagents such as SDS detergent and proteinase K, DNA is extracted with the phenol/chloroform solution by microcentrifuge. Then the purified DNA can be concentrated by ethanol precipitation.

---

*Address correspondence to Xing Chen; State Key Laboratory of Transducer Technology, Institute of Electronics, Chinese Academy of Sciences, Beijing 100190, China; phone and fax: +86-10-58887188; E-mail: xchen@mail.ie.ac.cn

Phenol/chloroform extraction is a liquid-liquid extraction technique. Typically, a 1:1 mixture of phenol and chloroform is added to an aqueous DNA sample in a microcentrifuge tube. The mixture is vigorously vortexed, and then centrifuged to enact phase separation. The upper, aqueous layer is carefully removed to a new tube, avoiding the phenol interface and then is subjected to two ether extractions to remove residual phenol. An equal volume of water-saturated ether is added to the tube with the mixture vortexed, and the tube is centrifuged to allow phase separation. The upper, ether layer is removed and discarded, including phenol droplets at the interface. After this extraction is repeated, the DNA is concentrated by ethanol precipitation.

Recently, commercially available silica column kits have been developed for nucleic acid purification. The supplied lysis methods and nucleic acid binding columns provide a fast (around 30 min), cost-effective way to purify nucleic acids. However, these methods still request centrifugation and manual pipetting by a technician.

Traditional methods and commercial kits for nucleic acid sample preparation are highly labor intensive and time consuming with multiple steps required to collect DNA or RNA from raw samples such as whole blood, urine, saliva, serum, tissues from biopsies, spinal fluid, and stool.

3. MICROFLUIDIC CHIPS FOR DNA PURIFICATION

With the development of the MEMS technology, macroscale sample preparation techniques for nucleic acid purification are modified and implemented in microfluidic chips, which can address several sample preparation challenges and simplify the sample preparation procedure.

The miniaturized nucleic acid purification methods typically reduce the total analysis time and cost by reducing the samples and reagents consumed and taking advantage of high reaction rates at the microscale. On-chip nucleic acid purification from whole blood samples or cell lines has been reported by several groups using different methods.

3.1. Solid Phase Extraction (SPE)

Solid Phase Extraction (SPE), which can be integrated integration into a microfluidic system, is one of the predominant methods currently used by microfluidic researchers. The chip-based DNA purification process, which relies on SPE, is achieved by the continuous flowing of the load buffer, the washing buffer, and the elution buffer, sequentially. The major advantage of the SPE method is that it avoids problems of physical and biochemical degradation of DNA during the purification process. However there is a problem facing the microfluidic chip-based SPE method, which is the limitation of the surface area of microfluidic chips, leading to low extraction efficiency, since the SPE method mainly relies on the interaction of DNA with solid phase matrix surface.

Kutter *et al.* [2] reported a micro SPE (µSPE) device for enriching coumarin C460 using the C18 wall coating method. However, this device was operated by an ultra-high voltage and can not address the capacity problem due to the limited surface area.

Laurell's group [3] reported a weir based silicon micro extraction chip packed with reversed phase beads with enhanced surface area for purification and enrichment of peptide mixtures containing urea for MALDI-TOF MS analysis. They also showed that trypsin immobilized beads could be used in this device for protein digestion, but did not report integration of the two steps. Furthermore, this same group [4] reported the second generation of the device by replacing the weir with a grid structure to confine the beads.

In addition to microbeads, it is also reported by Yu *et al.* [5] that monolithic porous polymers were prepared by photoinitiated polymerization within the channels of microfluidic devices and used for on-chip solid-phase extraction and preconcentration. In this monolithic microfluidic device, a concentration enhancement factor as high as 1650 on the dilute solutions of Coumarin 519 was reported under the optimal conditions. The performance in a more realistic application was then demonstrated with the enrichment of a hydrophobic tetrapeptide and green fluorescent proteins, where an increase in concentration by a factor as high as 10^3 was achieved.

The chip-based SPE method which can minimize sample loss, reduce analysis time and address contamination problems has also been widely applied for DNA purification, which is the indispensible step to prepare DNA samples for genetic analysis [6].

It is well known that the binding of DNA and RNA to silica, celite or glass powders is modulated by the presence of chaotropic agents such as sodium iodide, sodium perchloride, guanidine hydrochloride (GuHCl), and guanidine thiocyanate (GuSCN) [7-9]. Nucleic acid is able to bind with silica or glass fibers in high ionic strength solutions due to the decrease in the electrostatic repulsion. After washing with a non-polar solvent, DNA is eluted with a low ionic strength buffer. These procedures are well-known and commonly used in the commercial nucleic acid extraction kits.

As a preface to chip-based DNA extraction, Landers's group [1] established a µSPE system in a capillary packing with silica resin where PCR-suitable genomic DNA was directly extracted from human white blood cells, whole blood samples, and cells in culture, shown in Fig. **1**. In this paper silica beads were held in a polyethylene sleeve with two glass fiber frits and a silica capillary connected to serve as the inlet and outlet of this device. Silica beads were the solid phase matrix to apply the SPE method for nucleic acid extraction.

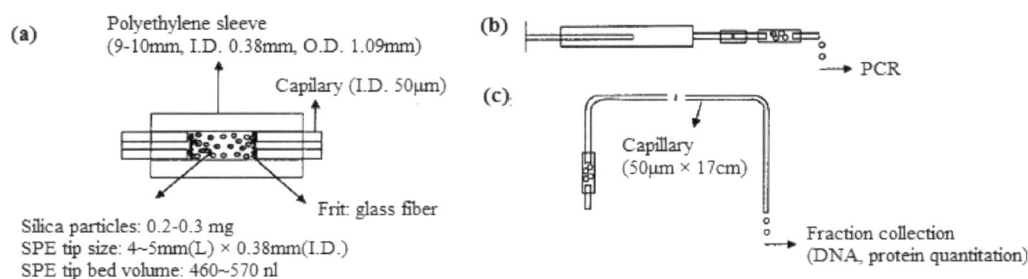

Figure 1: Schematic diagrams [1]. **(A)** Micro-solid-phase extraction device (µSPE device); **(B)** off-line mSPE; **(C)** on-line µSPE. Reprinted with permission from Academic Press. Copyright (2000).

In this design, DNA and RNA molecules were selectively absorbed by the surface of silica, celite, or glass fiber in the presence of high concentrations of chaotropic agents. Then the purification of DNA by silica-based solid-phase extraction was accomplished by the elution of nucleic acids with solutions of low ionic strength or even water, with the eluted DNA directly amenable to a variety of applications without additional concentration or precipitation steps.

Fluorescence assays were used to quantify the extracted DNA recovered from solid-phase resins by using PicoGreen as the fluorescence dye, while the polymerase chain reaction was used to evaluate the quality of the eluted DNA.

About 70% DNA was able to be directly recovered from white blood cells, while greater than 80% of the protein was removed with a 500-nl bed volume µSPE process that took less than 10 min. 10~30 ng DNA was extracted from white blood cells, cultured cancer cells, and even whole blood samples using 1 mg silica resin. Experiment results further demonstrate that the extracted DNA on the low microliter scale was suitable for direct PCR amplification and the miniaturized format as well as rapid time frame for DNA extraction was compatible with the fast electrophoresis on microfabricated chips.

This effort was followed by the work of the same group [12, 13] who exploited a sol-gel method to immobilize the bare silica beads, a sol-gel matrix or hybrid sol-gel/silica bead matrices within microchannels to maximize the extraction efficiency.

By comparing the extraction efficiencies of the three different solid phase matrix, hybrid sol-gel/silica bead matrix was found to yield the highest efficiency. Extraction efficiency with this matrix was averaged higher than 80 % with λ-phage DNA and even showed the efficiency higher than 90% for some microdevices. This provided a solid phase that yielded reasonably good extraction of DNA with good reproducibility and long term stability of the extraction bed.

In a more detailed study, the optimal loading buffer condition and flow rate were defined by using the hybrid sol-gel/silica bead matrix as the solid phase matrix [12, 13]. A higher DNA recovery was achieved at pH 6.1 rather than 7.6. And with the decrease in pH, the flow rate was increased correspondingly, which lead to the reduction in the extraction time from 25 min to less than 15 min. Under this optimized procedure, template genomic DNA from human whole blood was extracted on the microchip platform. The blood sample was mixed with the loading buffer before the mixture was pumped into the microfluidic chip. And the extracted genomic DNA from whole blood was successfully amplified. Both the microchip SPE procedure and a commercial microcentrifuge method were used to extract DNA from bacteria. The purification performance of the microchip device for bacterial DNA was shown to be comparable with the commercial microcentrifuge method. In the meanwhile, higher speed and lower sample consumption were claimed by the microfluidics based method.

However, this sol-gel/silica bead matrix exhibited a bonding and shrinkage problem between the microchannel wall and the matrix. To address this issue, a functional monolithic tetramethylorthosilicate (TMOS)-based sol-gel matrix with micro-pores was proposed and used to extract the DNA molecules [14]. The monolithic sol-gel bed was established in a microchip channel that provided enlarged surface area for DNA extraction with little flow-induced back pressure. The extraction efficiency of this system was about ~85% for λ,-phage DNA and about ~70% for genomic DNA from human blood. To demonstrate possible usage for clinical analysis, DNA molecules were extracted from bacteria and human cerebral spinal fluid, respectively, and amplified without any inhibition.

Furthermore, a photo-polymerized silica-based column was used as the solid phase matrix for DNA extraction, which was fabricated with the sol-gel solution consisting of a monomer solution of trimethoxysilylpropyl methacrylate (TMSPM) and photo-initiator [10].

A TMSPM-treated capillary was filled with the sol-gel solution and then exposed to UV light on a specific part of the microdevice, leaving the TMPSPM material only in the area exposed to UV.

Then the TMSPM monolith was modified with tetramethylorthosilicate (TMOS) using its enhanced hydrolysis properties under acid-catalyzed conditions. Upon acid hydrolysis, the hydrolyzed TMOS ($Si(OCH_3)_4$) molecules condensed with the unreacted siloxane/silanol groups present on the monolith surface, as displayed in (Fig. **2a**). The reacting TMOS molecules formed a continuous network of silicon dioxide, which increased the number of silica binding sites on the monolith surface.

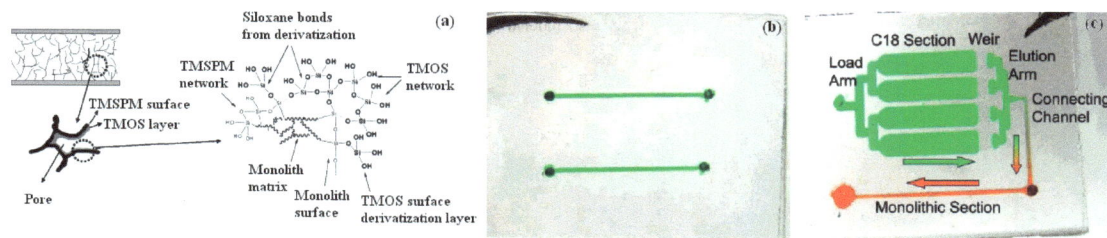

Figure 2: Schematic of TMOS derivatization on the TMSPM monolith surface **(a)** [10]; Dye-filled monolith DNA extraction microdevice **(b)** and integrated protein capture (green)/DNA extraction (red) in two-stage, dual-phase microdevice **(c)**. The channels are filled with dyes for better visualization of the protein and nucleic acid capture phase regions within the microchannel architecture. Arrow shows flow from stage 1 to stage 2. Chip dimension: 3 cm (length), 2.5 cm (width) [11]. Reprinted with permission from the American Chemical Society. Copyright (2006 and 2007).

The extraction efficiency was increased by using this modification procedure, in which about 85.6% DNA was extracted from the pre- purified genomic DNA, while only 59.9% DNA was extracted from the same sample mixed with blood [10].

A large mass of proteins present in blood was reported to have negative extraction efficiency of DNA by decreasing the value from higher than 80% to ~60%. To remove these proteins and therefore further enhance DNA purification, a two-stage microfluidic device was developed by also using photopolymerized monolith as the solid phase matrix in the glass microchips [11]. Two designs were investigated: a monolith DNA extraction chip (Fig. **2b**) and a two-stage DNA extraction chip (Fig. **2c**) consisting of a C18 (octadecyl) reversed-phase column for protein capture (stage 1) in series with a monolithic column for DNA extraction (stage 2). The glass microchips were fabricated through standard photolithography including wet etching and thermal bonding. The monolith matrix for DNA extraction was first photopolymerized in the microchannel and then modified by using TMOS, while the C18 beads were packed in the microchannels with weirs (Fig. **2c**). By combining the DNA extraction column with a C18 reversed-phase column with a high affinity for hydrophobic proteins, the two-stage microfludic chip showed a high protein removal rate and DNA extraction efficiency. From a 10-μL load of whole blood containing 350 ng of DNA, ~70% (1020 ± 45 ug) of proteins were retained, while a total of 240 ±2 ng of DNA was eluted from the second stage monolith, resulting in an overall extraction efficiency of 69 ± 1%.

Some of the critical researches in SPE focus on the location of optimized chaotropic salt, pH and elution profile for nucleic acids based on solid-phase extraction with silica beads [16]. Among four different salts (Guanidinium hydrochloride (GuHCl), guanidinium thiocyanate (GuSCN), sodium chloride (NaCl), and sodium perchlorate ($NaClO^4$)), GuSCN shows the highest adsorption on the silica surface. Usually, under low pH solutions, nucleic acids have more affinity with silica surfaces, but at pH between 6.0 and 8.0, adsorption isotherms show nearly the same affinity. Based on the elution profile, about 50% of the adsorbed nucleic acid has been collected in the first 5 ml of the elution buffer.

Christel *et al.* [17] and Cady *et al.* [15] used deep reactive ion etching or reactive ion etching on silicon to generate pillar structures with adequate surface area as solid phase matrix for DNA extraction.

One device with microfabricated pillars was etched to increase the surface area within the channel by 300–600% when the etch depth was varied from 20 to 50 um, shown in (Fig. **3**). DNA was selectively bound to these pillars in the presence of a chaotropic salt, followed by washing with ethanol and elution with a low-ionic strength buffer. For this device, the binding capacity of DNA was approximately 82ng/cm^2, in which about 10% of the bound DNA molecules were purified and recovered in the first 50 μl of elution buffer, and approximately 87% of the proteins were removed from a cell lysate.

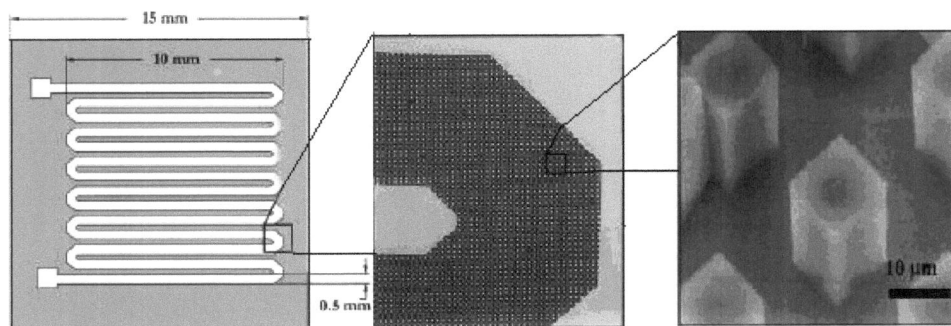

Figure 3: Schematic representation of channels containing microfabricated silica pillars [15]. The spacing between pillars and the pillar width was kept constant at 10μm, while the depth of the channels and height of the pillars could be adjusted between 20 and 50μm. Reprinted with permission from Elsevier. Copyright (2003).

Then an integrated microfluidic platform consisting of a DNA purification region with silicon pillar structures and a real-time PCR region was developed [18]. The whole procedure of DNA purification was achieved in approximately 15 min, while the real-time PCR for the detection was accomplished in 45 min for 10^4 *L. monocytogenes* cells and only 37 min for 10^7 cells.

With microfabricated silica structures on the wafer, the microfluidic device is able to make the extraction process simple and consistent. However the increase in surface area is limited and the potential problem of clogging could not be completely avoided.

Our group [19, 20] developed a µSPE chip using porous silicon with increased surface area as the solid phase matrix for DNA adsorption. SEM pictures of porous silicon channel are shown in (Fig. **4**). The pore size of porous rectangle channels anodized under the optimal conditions was determined in the range of 20 to 30 nm, and the total surface area was approximately $400m^2/g$ by using the BET technology. Thus the surface area to volume ratio of the porous microfluidic chip was approximately $300m^2/cm^3$, which is thousands of times higher than that of the non-porous counterpart. For the optimal µSPE chip, 49.5ng PCR-amplifiable DNA was extracted from one microlitre whole blood at the optimal condition within 15min of sample processing, which was approximately 2-fold higher than the performance of commercial kits.

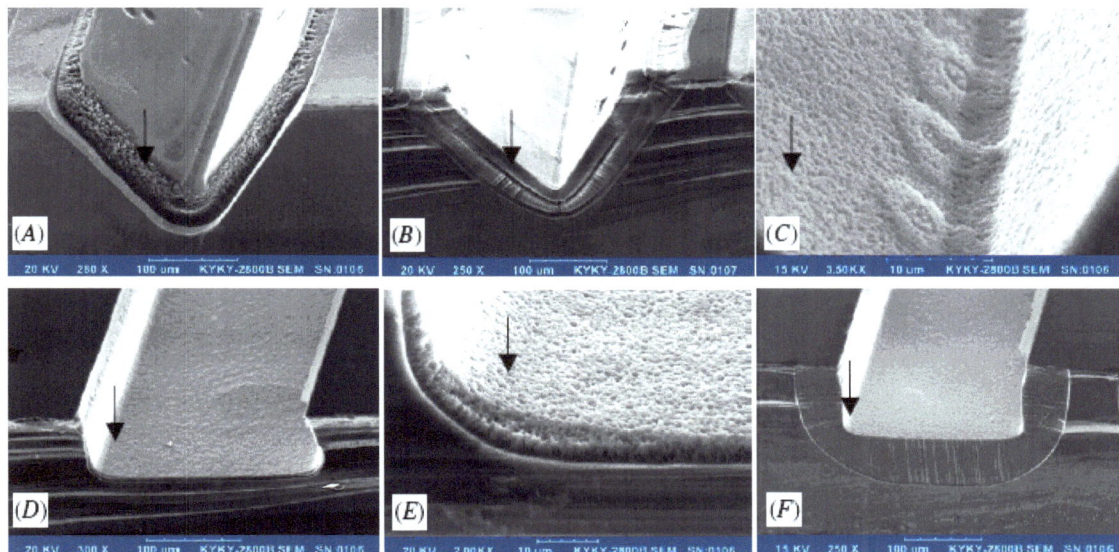

Figure 4: SEM micrographs of porous channels anodized in 30% HF electrolyte for 15 min [19-20]. (*A*) Porous V-type channels anodized at 30 mA/cm^2; (*B*) porous V-type channels anodized at 80 mA/cm^2; (*C*) higher magnification image of porous V-type channels anodized at 80 mA/cm^2; (*D*) porous rectangle channels anodized at 30 mA/cm^2; (*E*) higher magnification image of porous rectangle channels anodized at 30 mA/cm^2 and (*F*) porous rectangle channels anodized at 80 mA/cm^2.

Other studies employed amino-silane modified open channel microchips to extract DNA from blood with adsorption/desorption of DNA controlled by pH changes in the solvent [21]. Amine groups below neutral pH have a positive charge (causing negatively charged DNA to bind), which decreases above neutral pH. Cao *et al.* demonstrated a DNA SPE method based on changes in the charge of chitosan as the purification medium [22]. Chitosan has a cationic charge at pH 5 and is easily neutralized at pH 9. High density microfluidic channels coated with chitosan were fabricated and tested with lysed whole blood samples. DNA was captured at pH 5.0 through electrostatic interactions with the chitosan and eluted using the pH 9.1 Tris buffer. About 68% PCR-amplifiable human genomic DNA was extracted.

In addition, a microfluidic system was developed to purify mRNA from eukaryotic cells by packing with a UV-initiated methacrylate based Porous Polymer Monolith (PPM) [23]. The PPM polymerized and functionalized *in situ* provides a large surface area, good adhesion, and stability within a microchannel. Additionally, it was easy for the surface of a PPM to be modified with various functional groups. To extract mRNA from eukaryotic cells, the surface of the PPM was functionalized with oligo-dT. About 70% mRNA was extracted from high quality commercial poly-A-enriched RNA. And the capacity and purity of the extracted mRNA by using the PPM were equivalent to that of commercial kits.

A microfluidic device was also developed for the purification of total RNA from mammalian cells as small as 150 cells, or the equivalent of approximately 300 pg total RNA by packing silica beads into a microfluidic channel [24]. To increase the recovery ratio and purity of RNA samples, three silica bead columns with optimized silica beads and column pre-treatment were integrated with the microfluidic system. The RNA yield rate showed consistent increase as the number of cells was increased from 50 to 1000, and the total time for processing was about 50 min.

3.2. Solid Phase Reversible Immobilization (SPRI)

Solid phase reversible immobilization (SPRI) has been demonstrated to generate sequencing fragments that are free from template DNA, salts and excess dye-terminator products. SPRI is based on the selective immobilization of DNA onto carboxy-coated solid phase matrix [26-29]. Carboxylates and iron sites on the surface of the solid phase matrix act as a bioaffinity adsorbent for the targeted nucleic acids. Under conditions of high concentrations of poly(ethylene glycol), PEG, or tetra(ethylene glycol), TEG, and salts, DNA binds selectively to the surface of the solid phase matrix. Once bound, the solid phase matrix with the bound DNA is washed with ethanol to remove contaminants with the DNA subsequently released from the solid phase matrix in water or low ionic strength buffers yielding highly purified DNA. The SPRI method provides high DNA-binding capacity and assay reproducibility.

Xu *et al.* [25] demonstrated the purification of Sanger cycle sequencing fragments using the SPRI technology carried out in a UV-modified polycarbonate microchip. The chip was fabricated in polycarbonate (PC) that was stamped from a metal mold master using simple hot embossing, shown in (Fig. **5**). The microchannel containing microposts was used as the high surface area immobilization bed. By exposing the PC surface to UV radiation, a photooxidation reaction took place, which resulted in the formation of surface carboxylate groups.

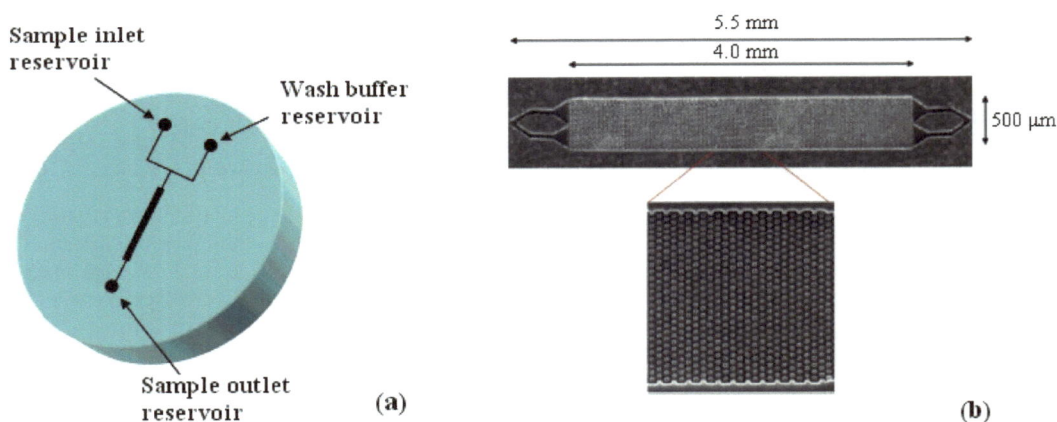

Figure 5: (a) Schematic diagram of the microfluidic device topology [25]. The microchannel used for SPRI was 500 μm in width, 50 μm in depth, and 4.0 mm in length. **(b)** Optical micrograph of PC SPRI capture-bed and its dimensions. Shown is the embossed piece in PC fabricated using the metal master. Reprinted with permission from the American Chemical Society. Copyright (2003)

DNA was precipitated onto the surface of the photoactivated PC microchannels. The loading density of DNAs to this activated surface was 3.9pmol/cm^2. Then the captured DNA molecules were subsequently released from the surface by incubation with ddH$_2$O.

Then Witek [30] used a photoactivated polycarbonate (PPC) microchip as a capture medium for gDNAs isolated from whole cell lysates. The recovery of DNA molecules following purification was estimated to be

85±5%. The immobilization and purification assay using this PPC microchip could be performed within ~25 min as follows: (i) DNA immobilization ~6 min, (ii) chip washout with ethanol ~10 min, and (iii) drying and gDNA desorption ~6 min.

As an extension of this research, a 96-well PC-SPRI microfluidic chip was developed by a high-throughput nucleic acid extraction system [31]. The loading capacity of nucleic acids from *E. coli* samples was 206 ng for gDNA and 165 ng for total RNA. 63% gDNA and 73% total RNA were extracted, respectively. PCR and reverse transcript PCR (RT-PCR) were successfully performed without any inhibition.

3.3. Magnetic Technologies

Magnetic technologies which can be easily integrated in microfluidic systems have also been exploited for the extraction of nucleic acids. Magnetic particles coated with silica or functionalized carboxyl groups have been used to extract DNA/RNA from biological samples [29, 32-34].

Magnetic beads with amine groups on the surface were used to increase the DNA binding affinity [21]. Bacterial magnetic particles (BMPs) modified with 3-[2-(2-aminoethylamino)-ethylamino]-propyltrimethoxysilane (AEEA) were used to absorb DNA by using electrostatic capture due to the increased amine yield. BMPs were magnetic particles synthesized by magnetic bacteria, which can be used for free floating in solution and then collected in one location using a magnetic field.

DNA was extracted from whole blood samples using aminosilane-modified BMPs, shown in (Fig. **6**). Whole blood was mixed with a lysis buffer, and incubated at 56°C for 20 min. Aminosilane-modified BMPs were added to the blood lysate. DNA–BMP complexes were collected magnetically and the supernatant were removed. The complexes were washed five times with the TE buffer. Finally DNA was released from the BMP complexes after the DNA–BMP complexes were incubated in an elution buffer for 20 min at 80°C. 95% DNA molecules were recovered from the aminosilane-modified BMPs.

Functionalized Magnetic Particles (MPs) coated with sequence-specific probes were also used to extract *E. Coli* genomic DNA in a microreactor reported by Yeung [35]. The silicon/glass-based chip was fabricated with a platinum heater and temperature sensors for thermal cell lysis and genomic DNA denaturation to provide an accessible DNA template for MPs to capture by using the biotinylated DNA probes. The denatured genomic DNA hybridized with a biotinylated DNA probe on the particle and then the hybridized gDNA was separated from the remainder of the cellular components in a wash step. The captured gDNA was released by heating the silicon/glass-based microreactor without a specific elution buffer. Using this technique, one can extract species-specific DNA, but a few extra steps were required to functionalize the magnetic particles and more time was needed to effectively hybridize the target with the probe.

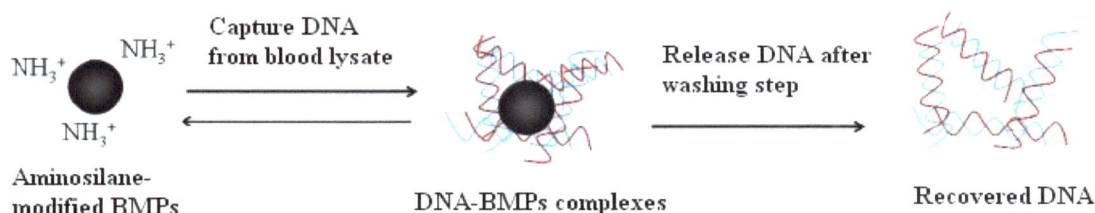

Figure 6: Schematic of how magnetic microparticles functionalized with aminosilane are used to capture from the blood lysate and release from modified BMPs. acid from blood samples [21]. Reprinted with permission from Wiley-VCH Verlag GmbH & Co. KGaA. Copyright (2006).

A magnetic bead-based microfluidic platform integrating three major modules for rapid leukocyte purification, genomic DNA (gDNA) extraction and fast analysis of genetic gene was developed by adopting microfluidic technologies with functionalized magnetic beads [36]. Magnetic beads conjugated with $CD_{15/45}$ antibodies were used to isolate and concentrate leukocytes from human whole blood samples.

Then surface-charge switchable, DNA-specific, magnetic beads were utilized to extract and purify DNA from the blood lysate. This system allowed cell lysis and DNA extraction within 10 min and was able to generate a high yield and good purification of DNA samples in an automated fashion.

A totally integrated pathogen DNA sample preparation system was developed using a microfluidic Compact Disk (CD) platform [37]. Cell lysis is accomplished rapidly by using a laser system without any lysis buffers and gDNA is extracted using biotinylated magnetic beads. The extraction efficiency produced results similar to that obtained with commercial DNA preparation kits.

Harrison's group [38] demonstrated mRNA capture in a Y-shaped microfluidic chip device using paramagnetic oligo-dT beads and magnetic trapping to capture, and then release the beads. Using Drosophila melanogaster (fruit fly) DNA, they were able to capture ~2.8 ng mRNA from 0.85 microgram and 34 ng mRNA from10g tRNA. The amount of mRNA was high enough for cDNA (complementary DNA) library construction using RT-PCR [38, 39].

A drawback to use magnetic particles in microfluidic systems is the requirement of appropriate micro-manipulating systems to control the magnetic field.

3.4. Liquid/Liquid Extraction

Liquid/Liquid Extraction (LLE) is another alternative method for extraction and purification, which has been used in conventional sample pre-treatment where a compound (or a mixture of compounds) is transferred from one liquid phase to another. Under the micro scale, the high surface-to-volume ratios and short diffusion distances, combined with laminar flow conditions, offer the possibility of performing chip-based LLE systems.

Kitamori and colleagues [40, 41] were pioneers in realizing the merits of the microscale LLE and have reported several microfabricated devices for LLE. One of the microfluidic devices is a variation of an H-filter design, fabricated in quartz. For this given design, Fe(II) was introduced in an aqueous stream and trioctylmethylammonium chloride in an organic (chloroform) stream, and then the ion-pair product was extracted in the organic phase in less than 45s, representing an order of magnitude improvement over conventional extraction times in separation funnels. LLE was achieved by contacting fluidic streams at constricted openings between distinct channels. The approach is attractive since flows can be separated naturally as the channels diverge.

Reddy and Zahn [42] generated dual inlet and three inlet microfluidic systems based on organic–aqueous liquid (phenol) extraction, which could be used for purifying DNA directly from cells. Microchannels were fabricated by using hydrofluoric acid wet etching of glass with a chrome–gold masking layer. The phenol and water phases were infused into the microchannels, shown in (Fig. 7). The cell components naturally distributed themselves into the two fluid phases in order to minimize interaction energies of the biological components with the surrounding solvents. The membrane components and protein partitioned to the interface between the organic and aqueous phases while the DNA stays in the aqueous phase. The aqueous phase was then removed with a purified DNA sample.

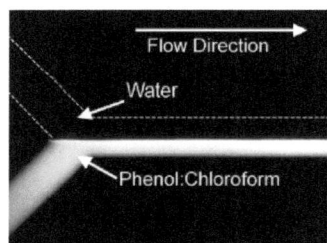

Figure 7: Co-infusion of water and phenol:chloroform solution showing a stable stratified flow profile [42]. The phenol is dyed with a lipophilic dye and is bright under epifluorescent microscopy. Reprinted with permission from Elsevier. Copyright (2005).

4. PRINCIPLE OF SOLID PHASE DNA EXTRACTION

The majority of DNA purification microfluidic chips use the principle of solid phase extraction for DNA purification, which relies on the adsorption of DNA onto a solid surface, followed by the washing of the impurities such as proteins, and finally the pure DNA is eluted in the aqueous solution. The mechanism of DNA adsorption on a solid surface under high ionic strength chaotropic conditions elucidated by Melzak [43] is that the adsorption of highly charged duplex DNA to hydrophilic negatively charged silica is controlled by three competing effects: weak electrostatic repulsion forces, dehydration and hydrogen bond formation. The high ionic strength serves to shield the negative surface, reducing the electrostatic repulsion between the negative DNA and the surface of the silica, while the chaotropic salt dehydrates the silica surface and DNA, thus promoting hydrogen bonding between the DNA molecules and the protonated silanol groups. These two factors combine to allow DNA to adsorb onto silica surfaces.

The high ionic strength serves to shield the negative surface, reducing the electrostatic repulsion between the negative DNA and the surface of the solid phase matrix, such as silica, glass beads and silica resin, while the high concentration binding salt dehydrates the solid phase matrix surface and DNA, thereby promoting hydrogen bonding between the DNA molecules and the protonated silanol groups. The adsorption process is reversible. In the presence of high pH and low concentration binding salt, free water molecules could hydrate the matrix surface and DNA again, thus resulting in the hydrogen bond breaking, while the low ionic strength enhances the negative surface, increasing the electrostatic repulsion between the DNA and the matrix surface. Finally, these factors combine to allow the DNA to desorb from the matrix, shown in (Fig. **8**). Perchloride salt, guanidine, potassium iodide, *etc.* have been used as the bonding salts and the concentrations are about 4~6 M, with pH 6.4~6.7. In the same condition, the surface area of the matrix significantly affects the extraction efficiency of the DNA, and thus the extraction efficiency improves with the increasing surface area-to-volume [13, 17].

5. CASE STUDY I: SOLID PHASE MATRIX FOR DNA EXTRACTION

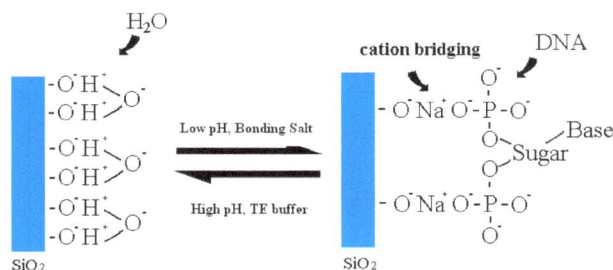

Figure 8: Schematic of the DNA absorption and desorption process.

It is well known that Porous Silicon (PS) has the advantage of large specific surface area-to-volume (hundreds of meter squares per cubic centimeter), which can be used in the specific application fields requesting high effective interface area. Since porous silicon was first reported in 1950s [44, 45] by means of electrochemical etching of silicon wafer surfaces, porous silicon has been widely acknowledged due to its unique properties, such as efficient electroluminescence, high resistance, high specific surface area and so on. On this porous silicon surface, the pore sizes and surface morphologies can be controlled by electrochemical etching conditions such as current densities, HF concentrations and etching time [46, 47]. Porous silicon has been used in various fields of solar cells [48-50], Radio Frequency (RF) [51, 52] and (bio)chemical sensors [53-55]. Furthermore since the technology for porous silicon fabrication is compatible with standard microelectronic and MEMS techniques, porous silicon as a biocompatible solid supports has been used to absorb enzyme, protein and other biologic molecules in fields of the enzyme micro reactors [56, 57], chromatography [58], and antibody micro arrays [59].

This example aims to illustrate the preparation, characterization and optimization of the porous silicon based solid phase matrix for DNA absorption [20, 60].

5.1. Fabrication of Porous Silicon

In this specific device, the porous matrix was fabricated on the surface of the inner wall of microfluidic channels by MEMS technologies and the electrochemical etching technology. The process used to fabricate the chip with porous silicon matrix was as follows. Fabrication started with the n-type 0.01-0.1Ω·cm silicon wafers of (100) crystal orientation, which are double-side polished with approximately 350μm in thickness. The silicon wafers were first coated with 0.3μm thick of silicon nitride, as a mask for the anisotropical etching and the subsequent electrochemical etching. Then the wafers were spin coated with a positive photoresist (AZ1500) and patterned. After the exposed photoresist was developed, the exposed silicon nitride on the wafers was removed by plasma etching to produce a bare silicon upper surface. And then the wafers were etched in 33% potassium hydroxide (KOH) by weight for 3h to produce microchannels of about 140μm deep. Another alternative fabrication process was that the wafers were etched in a deep reaction ion etcher (Adixen, AMS100). Then a surface-enlarging porous silicon layer was obtained on the internal walls of microchannels by electrochemical etching the patterned silicon in an electrolyte which was a mixture of hydrofluoric acid (HF) and ethanol. Different concentrations of HF and different current densities were investigated with various etching time. After the process of electrochemical etching, the silicon wafers were thoroughly rinsed in distilled water. Then the wafers were put into a furnace where the temperature was increased up to 500°C from room temperature within half an hour and then kept for one more hour, to cover the porous silicon layer with a silicon dioxide layer.

In the meanwhile, microfluidic channels with micropillar arrays were fabricated by etching silicon wafers in a deep reaction ion etcher (Adixen, AMS100) without any electrochemical modifications. These microfluidic channels were used as control devices to further evaluate the performance of porous silicon based microchannels.

5.2. Characterization and Optimization of PS

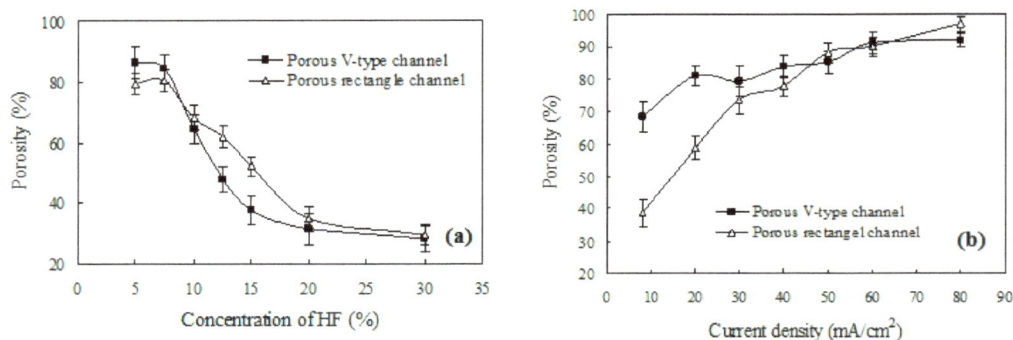

Figure 9: (a) Effect of HF concentration on the porosity of porous channels. The porous silicon was anodized at 30mA/cm^2 for 15min. **(b)** Effect of current density on porosity of porous channels. The porous silicon was anodized in 10% HF electrolyte for 30min.

Porous silicon has a huge surface area, which has been used as matrix for biochemical analysis [61, 62]. Porous silicon is fabricated by anodizing Si in HF electrolyte, and a variety of morphologies can be obtained depending on anodization conditions, wafer types with dopant levels. It is also found that surface morphologies of porous silicon on the internal walls of channels can be strongly affected by the shape of the channels.

Porosity of porous channel was firstly investigated as a function of a group of preparation parameters. The porosity is defined as the faction of voids within the porous silicon layer and can be easily determined by weight measurements. The porosity of porous channels was determined gravimetrically after dissolving porous silicon by 3% KOH by weight. Silicon wafer was weighed before anodization (m_0), just after anodization (m_1), and after a rapid dissolution of the whole porous layer in a 3% KOH solution (m_2). The porosity (P) was thus calculated by the following equation:

$$P(\%) = (m_0 - m_1)/(m_0 - m_2) \tag{1}$$

Moreover the practical thickness (d) of the porous rectangle channels was measured by a surface profiler (Tencor, Alpha-step 500) after dissolving porous silicon using 3% KOH by weight. The theoretical thickness (D) of the porous rectangle channels was also determined by weight measurements, which was given by the following equation:

$$D = (m_0 - m_2)/\rho \cdot S \tag{2}$$

Where S is the internal surface area of the rectangle channel and ρ is the density of Si.

5.2.1. Effect of HF Concentration

To obtain insight into the effect of HF concentrations on porosity of porous silicon, porous V-type channels and porous rectangle channels were prepared with HF concentrations ranging from 5 to 30% by volume, while the current density and etching time were kept constant. Porosity of porous channels was calculated by the equation (1). As shown in (Fig. **9a**), the porosity of both porous channels increased with the decrease in HF concentration. With the decrease in HF concentration, the anodic etching became weaker and the chemical dissolution was stronger, which lead to higher porosity. When the concentration of HF was higher than 20%, the porosity of both porous channels was decreased to approximately 30% and showed the independence of channel shapes.

The morphologies of both porous channels were also investigated in the 30% HF electrolyte at different current densities. SEM pictures of both porous channels are shown in (Fig. **4**), which shows that the surface morphologies of both porous channels were insensitive to the shape of channels and current densities when the HF concentration is high enough (*e.g.*, 30%). Moreover the thickness of the porous layer was noticed to increase with the increase in the current density because of the increase of the electric field.

5.2.2. Effect of Current Densities

The effect of the current density on porosity of porous channels was investigated at different current densities from 8 to 80mA/cm^2 when the HF concentration and etching time were fixed. The result is shown in (Fig. **9b**). In the current study, there was an increase of porosity with the increase in the current density, which agreed well with the results of Ryel Kwon [63]. Since the electric field was increased at higher current densities, much more bulk silicon was dissolved through direct transfer of Si atoms into the silicon/hydrofluoric acid electrolyte interface, resulting in higher porosity.

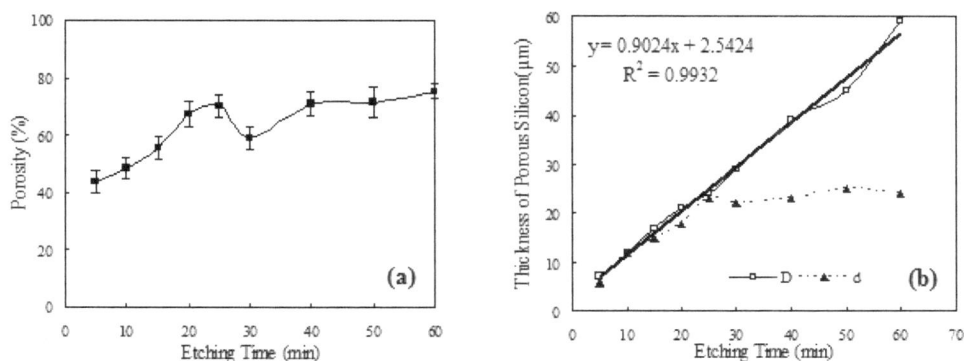

Figure 10: Effect of etching time on thickness of the porous silicon layer. The porous silicon was anodized at 20mA/cm^2 in 10% HF.

The surface morphology of porous channels was independent of the current density and the shape of channels in 30% HF electrolyte. However, this relationship was quite different when the HF concentration was lower.

When the current density was 20mA/cm^2, surface morphologies of porous V-type channels appeared to be non-uniform, whereas the rectangle one showed a homogenous spongy structure. When the current density was increased to 80mA/cm^2, some porous silicon in V-type channels became to peel off with unstable surface morphology, whereas porous rectangle channels demonstrated more stable structures like a dried lake texture. A simple model for pore formation in silicon is proposed, which is based on the transportation of minority charge carriers, *i.e.,* holes, through the bulk silicon to the silicon/hydrofluoric acid electrolyte interface. The migration of these charge carriers in the bulk silicon is mainly diffusion driven, but in the vicinity of the concave structures the charge carrier transport is controlled by the electrostatic field imposed on the silicon structure during the anodization process as described Laurell [64]. Thus, the pore becomes larger and deeper in concave corners which results in the uneven stress at the interface of the bulk silicon and the porous silicon layer. For porous V-type channels, this effect tends to be more obvious. As a whole, surface morphologies of porous rectangle channels are more homogeneous and stable than that of porous V-type channels.

5.2.3. Effect of Etching Time

To determine the effect of etching time on porosity and thickness of the porous silicon layer, porous rectangle channels were fabricated for different etching time, ranging from 5 to 60 min, with the current density and the HF concentration unchanged. The theoretical thickness of the porous silicon layer (D) was calculated according to the equation (2) and the practical thickness (d) was obtained by a surface profiler. The thickness was measured after anodization (d1) and after a rapid dissolution of the whole porous layer in 3% KOH solution (d2). The practical thickness of the porous silicon layer (d) was thus the difference between d1 and d2. The porosity of porous rectangle channels was noticed to first increase with etching time in the first 25 min, with porosity of 70.2%, and then decrease to 58.7% in the interval of 25-30 min (Fig. **10a**). This reduction might be due to the fact that a significant portion of porous silicon got dissolved chemically during that period. Finally, the thickness increased again slowly in the interval of 30-60 min in accordance with the process of anodic etching.

The effect of etching time on the thickness of porous silicon is shown in (Fig. **10b**). The experimentally obtained thickness was consistent with the theoretical calculation in the first 25 min, which showed linear increase with the etching time stated by Clicq [58]. However the experimental obtained thickness gradually deviated from the theoretical ones after 30 min of anodization. Possible reasons are as follows. The chemical dissolution of the porous silicon layer is weak that it might be negligible within a short period of the etching. But when the etching time is long enough, it takes much longer for Si atoms to reach the solution to form pores while the mass of chemically dissolution of porous silicon increases. As a result, the experimental obtained thickness was much lower than the theoretical calculations in a longer period of silicon etching.

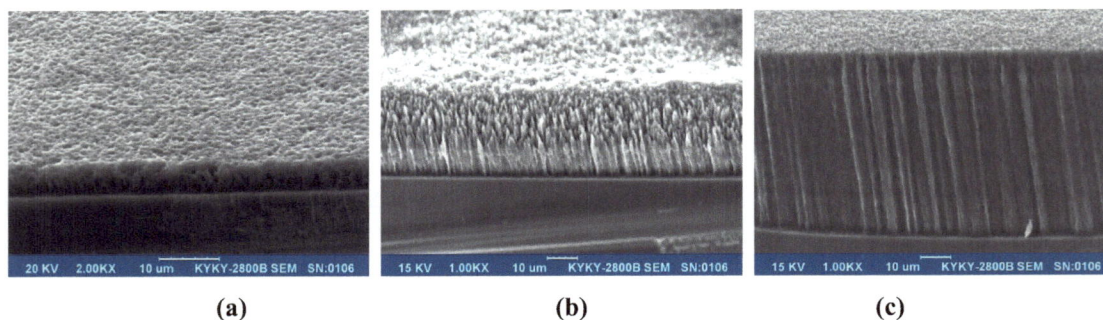

(a)   (b)   (c)

Figure 11: Cross-sectional SEM images of the porous silicon layer on the internal walls of the rectangle channels anodized in an ethanol (95%)/HF (48%) solution (mixing ratio 2:3) at 8 mA/cm^2 **(a)**, at 30 mA/cm^2 **(b)** and at 80 mA/cm^2 **(c)**. The etching time is 15min.

5.2.4. Optimization of Anodization Conditions

Porous silicon cannot be used for DNA purification until it is oxidized to form porous silicon dioxide. For a higher porosity of porous silicon, the thermal mismatch stresses occur and the thickness of the oxidized layer is probably changed, whereas for a lower porosity, it is extremely difficult for the porous silicon layer to be completely oxidized. Based on the previous publication [65], porosity of 46-55% is the optimized porosity for achieving a complete oxidation.

Anodization experiments record a thickness increase of the porous layers from several microns to several tens of microns as a function of the increase in current density (Fig. **11**). Although Clicq D. [58] reported that the thickness of porous silicon increased linearly with etching time, the upper porous silicon layer would be peeled off easily due to the chemical dissolution effect when etching time is too long. In order to make the porous layer more homogeneous, the etching time should be decreased. Therefore to obtain more uniform morphologies, higher porosity and more stable structures for easy oxidization, the current density of 20 mA/cm^2, the HF concentration of 10% and the etching time of 15min were chosen as the optimal condition to fabricate porous channels with the porosity of 55.6%. Under these anodizaiton conditions, homogenous porous silicon was fabricated and pores were located at the silicon interface and oriented perpendicularly to the internal surface of the microchannels.

According to IUPAC guidelines the porous silicon is classified depending on pore sizes as microporous (<2 nm), mesoporous (2–50 nm) and macroporous (>50 nm) [66]. Mesoporous silicon has a huge surface area with moderate pore size, which was used as a matrix for immobilizing biological macromolecules [67, 68]. Accurate determination of the surface area of porous matrix is usually given by the analysis of adsorption isotherms of gases at low temperature (BET technique). By using BET nitrogen adsorption experiments, the pore size distribution for the fabricated porous silicon was characterized as in the range of 10 to 30nm, which belonged to mesoporous silicon under the fabrication conditions of 20mA/cm^2 in 10% HF electrolyte for 15min. The yield BET surface area of the porous silicon was about 300m^2/g, and thus the total surface area of the chip was about 0.36m^2 with about 0.00127g porous silicon.

5.3. Comparision of Various Solid Phase Matrix

Microchannels (25cm long and 200μm wide) were fabricated on a silicon wafer based on BioMEMS technologies, and the walls of the channels were directly used as the solid phase matrix for DNA extraction without packing solid phase matrix. In this example, five kinds of different solid phase matrixes were fabricated and compared, which are V-type channels (1#), rectangle channels (2#), silicon pillars (3#), porous silicon on the wall of V-type channels (4#) and porous silicon on the wall of rectangle channels (5#). The inner surface of channels was used as the solid phase matrix in (1#) and (2#), while the inner surface of the rectangle channels and the surface of micro-pillars were used as the matrix in (3#). The porous layers on the walls of the V-type channels and the rectangle channels were used as the matrix in (4#) and (5#).

Table 1: Characteristics of solid phase matrix and key fabrication procedures

	Solid phase matrix	Depth of channel (μm)	Surface area * (cm^2)	Key technology
1#	V-type channel	100 141.42	0.759 0.865	Anisotropic etching
2#	rectangle channel	100 150	1 1.25	DRIE
3#	silicon micro pillar **	100 150	2.86 4.09	Optical lithography DRIE
4#	porous silicon V-type channel ***	100	1783	Electrochemical etching
5#	porous silicon rectangle channel ***	150	3600	DRIE Electrochemical etching

*: The length of the channel is 25cm. **: The diameter of illars is 20μm. The gap of pillars array is 20μm. ***: The surface area of porous silicon channel was obtained by BET (Brunauer, Emmet, and Teller) nitrogen adsorption analysis.

Five types of solid phase matrices are listed and compared in Table **1**. From the point of microfabrication, if silicon wafers are etched in an anisotropic solution, the etching rate is highly dependent on the crystal direction. When the silicon wafers of (100) crystal orientation are etched, the etching section is oblique cone shape, and finally the etching cross section is a V shape. Based on theoretical calculation, the ultimate etching depth of channels with 200 µm wide is 141.42µm. It is easy to fabricate V-type channels by the anisotropic etching technology without any special equipment. However the V-type channel suffers from the limitation in the surface area. The rectangle channel with higher depth and silicon micro pillar arrays can be fabricated by the DRIE technology with an increase in surface area. Since DRIE is requested, there is an increase in fabrication difficulties. In the meanwhile, the surface area of the porous silicon solid phase matrix fabricated by the electrochemical technology can record thousands of times increase without the need of DRIE. Since the electrochemical technology is compatible with BioMEMS technologies, where the morphology, porosity, and depth of porous silicon can be precisely controlled by the electrochemical etching conditions with high fabrication repeatability [69-70], we selected the porous silicon on the wall of channels as the matrix for DNA extraction followed by the oxidation of the porous silicon layer.

5.4. Biocompatibility of the Porous Matrix

The DNA purification process employed here utilizes the adsorption of DNA onto the solid phase matrix under high ionic strength chaotropic conditions, followed by washing the impurities such as proteins. Finally the desorbed DNA was eluted in an aqueous solution. The mechanism of DNA adsorption on the solid surface elucidated by Melzak [43] is that the adsorption of highly charged duplex DNA to hydrophilic negatively charged silica is controlled by three competing effects: weak electrostatic repulsion forces, dehydration and hydrogen bond formation. However, the surface of the bulk silicon is hydrophobic in the process of pore-making [71], which makes the adsorption of nucleic acids and the flowing of fluid in channels impossible. To overcome this problem, the thermal oxidization of the porous silicon was used to form amounts of silanol groups which can interact with nucleic acids and behave as a hydrophilic surface. (Fig. **12**) shows the droplet on the porous substrate before and after the thermal oxidation process. Static contact angles on substrates before and after thermal oxidation process were about 75° and about 20°, respectively. These results clearly indicate that the surface of porous silicon after thermal oxidation was more hydrophilic than the surface without thermal oxidation. Moreover, the porous SiO_2 layer formed by thermal oxidation also showed better passivation properties, stability and chemical inertness than the bare porous silicon layer.

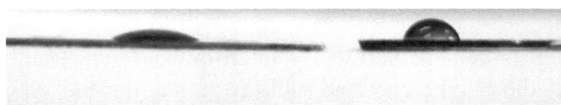

Figure 12: Pictures of water droplets (1µL) on porous silicon substrates before (right) and after (left) the thermal oxidation process.

6. CASE STUDY II: DNA EXTRACTION USING A MICROFLUIDIC SPE CHIP

Various silicon microfluidic chips with porous and non-porous solid phase matrix layers on the walls of the channels have been used to extract DNA from whole rat blood, culture cells and other samples [20, 72-74].

6.1. Experiment

6.1.1. Fabrication of Microchips with Porous Matrix

The microfluidic chip with a porous matrix was fabricated by MEMS technologies and the electrochemical etching technology. As shown in Fig. **13**, the chip consists of a glass cover and a silicon substrate including a tortuous microchannel with a porous layer on top of it. The SPE microfluidic chip with porous v-type channel was fabricated as follows.

N-type 0.01-0.1 Ω/cm silicon wafers of (100) crystal orientation were coated with silicon nitride, as a mask for anisotropical etching and electrochemical etching, after the wafers were cleaned thoroughly. Then the wafers were spin coated with a positive photoresist (BN303) and patterned. After the exposed photoresist was developed, the exposed silicon nitride on the wafers was removed by plasma etching. Then the wafers were etched in 33% potassium hydroxide (KOH) by weight to obtain a V-type channel. A porous silicon layer on the internal walls of channels was obtained by the electrochemical method. Then the wafers were thermally oxidized at 500°C. Finally, a Corning Pyrex#7740 glass cover was anodically bonded to the silicon wafer to form a close channel by a bonder (Suss, SB6).

Figure 13: Pictures of the SPE microfluidic chip with porous matrix including an inlet and an outlet.

The SPE microfluidic chip with a porous rectangle channel was fabricated as follows. The silicon wafer was first deposited with 0.3 μm thick of silicon nitride (Si_3N_4) as the mask for fabricating the porous silicon, and then with 0.3 μm aluminum as the mask for deep reaction ion etching (DRIE). After photolithograph, the wafer was etched in a deep reaction ion etcher (Adixen, AMS100) to obtain a rectangle channel. A porous silicon layer on the internal walls of channels was obtained by the electrochemical method. After that, the silicon wafer was thoroughly rinsed in distilled water and oxidized at 1050°C for 1 h. Finally a glass cover was anodically bonded to the silicon wafer.

6.1.2. DNA Extraction from Whole Blood Samples

Microchips were washed with HNO_3 and the TE buffer for 5min prior to each extraction experiment. The typical experimental procedure is shown as follows: the buffer solution with 10μL whole blood and 4μL Triton X-100 and 100μL A were thoroughly mixed and then incubated in a water bath at 40~55°C for 5min for lyzing cells and then pumped into the chip at the flow rate of 10~15μL/min for 8~10min. After the load buffer flowed out from the microchip, the solution was collected and then marked as S1. The washing buffer with 30μL B solution and 30μL ethanol was thoroughly mixed and pumped through the chip at the flow rate of 20~25μL/min for 2.5~3min. After this washing buffer was flowed out, this solution was collected and then marked as S2. Then the 70% ethanol solution (by volume) was pumped through the chip at the flow rate of 20~25μL/min for another 2.5~3min. After this washing solution flowed out, every 30μL solution was collected and then marked as S3 and as S4. A 20μL TE buffer was introduced in the chip for 10min incubation and then flowed out. This TE buffer was collected and marked as S5. Then a 100μL TE buffer was continuously pumped through the chip at the flow rate of 5~10μL/min for 10~20min. Every 20μL TE buffer was collected and marked as S6~S11, respectively.

6.1.3. DNA Detection

DNA collected in the load, wash, elution solutions was quantified by using the SYBR Green I dye tagging DNA. A real-time quantitative PCR detecting system was chosen to measure the intensity of fluorescence and then the DNA concentration was calculated by using calibration curves, which were generated using lambda DNA. 10μL solutions from S1~S11 were added with 4μL SYBR Green-I (v:v = 1:1 000) and 11μL TE buffer, respectively. Then these solutions were marked as S'1~S'11, respectively.

The DNA extracted from whole blood was quantified by Polymerase Chain Reaction (PCR) and gel electrophoresis. 5μL DNA extracted by the microchip was used as the temple of PCR to amplify the 203-bp fragment of Gapd gene, in addition with positive and negative controls. The PCR reactions consisted of 2.5μL of the standard 10×PCR buffer, 100μmol of dATP, dGTP, dCTP, and dTTP, 2.5 units of Taq polymerase, 50nmol/L of each primer, and 5μL of the initial collected fraction, in a total volume of 25μL. These reactions were cycled under the following conditions: 95°C denaturation for 5min, 35 cycles of 94°C for 1min, 68°C for 1min, 72°C for 1min, followed by a 10min extension at 72°C. The PCR products were then tested in gel electrophoresis.

6.2. DNA Extraction by Using the V-Type Porous Microchip

6.2.1. Yield of the Extracted DNA

The DNA purification process employed here utilizes adsorption of DNA onto a solid phase matrix of silicon dioxide or glass under high ionic strength chaotropic conditions and low pH, followed by the desorption of DNA from the matrix under low ionic strength chaotropic conditions and high pH. The yield of extracted DNA is not only affected by the type of salt, the solution concentration, and the pH of the solution, but also by the specific surface area-to-volume and surface characteristics of the solid phase matrix. According to the literatures [1, 15, 17] and our own experimental observations, 6mol/L GuHCl at pH 6.7 was used as the bonding salt for DNA extraction.

The non-porous silicon dioxide microchip for extracting DNA was fabricated by MEMS technologies, while the porous silicon microchip was fabricated by anisotropical etching and then oxidation at the room temperature and higher temperatures for DNA extraction. Under the same conditions, these two kinds of microchips were used to extract genomic DNA from the whole blood samples. The yield of extracted DNA was measured by the fluorescence reagent of SYBR Green I. The intensity of the fluorescence in the presence of double stranded DNA is much higher than that without the presence of double-stranded DNA. Moreover, the intensity of the fluorescence is linearly increased with the concentration of DNA. Thus, the concentration of DNA could be calculated by means of a calibration curve, which was generated using lambda DNA. Experimental results show that 15.7 ng genomic DNA was extracted from one microlitre whole blood by using the non-porous silicon dioxide microchip (chip #1), while 9.2 ng genomic DNA was extracted by using the oxidized porous silicon microchip (chip #2) oxidized at room temperature. The yield of DNA extraction was affected by the temperature of oxidization and approximately 134 ng genomic DNA was extracted by using the oxidized porous silicon microchip (chip #3) with oxidation at 400°C, whereas approximately 25 ng of genomic DNA was extracted by using the oxidized porous silicon microchip (chip #4) with oxidation at 500−700°C. Since the silicon wafer was significantly misshaped during the process of oxidation at high temperature, these devices after oxidation cannot bond to the cover plate and therefore cannot be used for DNA extraction with the yield of extracted DNA zero by using this microchip (chip #5). Compared to non-porous silicon dioxide, oxidized porous silicon has a much higher surface area and much more silanol groups for adsorbing DNA, leading to the improved yield of DNA extraction. When the porous silicon was oxidized at the room temperature, the oxidized porous silicon layer was thinner and the quantity of Si−O bonds was lower [75]. Thus the total number of silanol groups was lower and the ability of DNA adsorption was weaker. In summary, the oxidized porous silicon microchip with oxidation at 500°C was used to extract DNA in the following experiments due to the best performance.

6.2.2. DNA Extraction from Rat Peripheral Blood

Genomic DNA was extracted by using the oxidized porous silicon microchip oxidized at 500°C, and the fluorescence intensity of all of the collected samples was quantified by using the fluorescence method. The concentration of DNA in all of these collected samples was obtained by means of comparing to the calibration curve, which was prepared using lambda DNA. The results are shown in (Fig. **14**). The concentration of DNA in sample S′1 was 0 ng, which indicates that the majority of genomic DNA was adsorbed onto the oxidized porous silicon during the load step. The concentration of DNA in samples S′2~S′4 which represents the collected solutions during the washing step was very low, which indicates that only a little amount of DNA was desorbed during the washing step. Samples S′5~S′11 represent the eluted solutions for desorbing DNA from the oxidized porous silicon.

From (Fig. **14**), the value of S′5 was only 22.4ng which indicates that the desorbed DNA during the first elution step, and the value of S′6 was 39.5 ng, the highest point, which indicates the desorbed DNA during the second elution step. The highest point of DNA desorption doesn't occur after the first elution step, which indicates that the procession of DNA desorption from oxidized porous silicon was slow. Overall, 97.6% genomic DNA was desorbed from oxidized porous silicon after five-time elution. Roughly 24 ng genomic DNA was extracted from every microlitre whole blood by using the oxidized porous silicon microchip, with comparable performance with commercial kits which can extract 20~30 ng DNA from one microlitre whole blood.

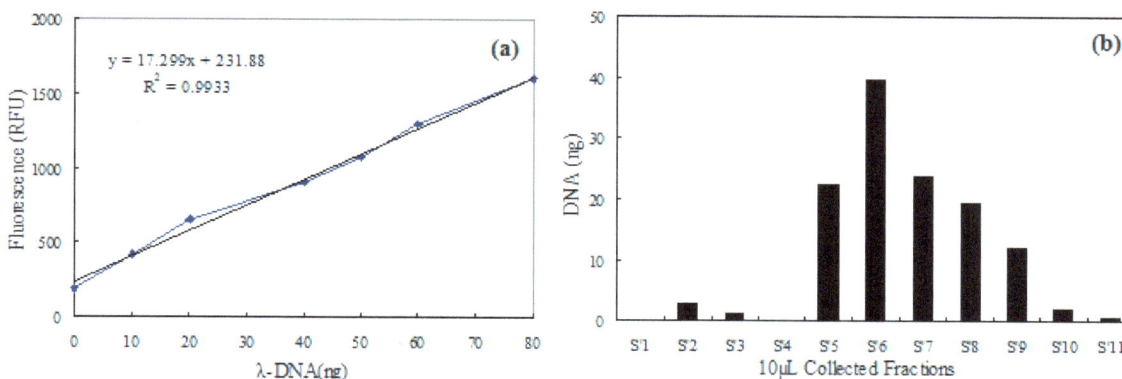

Figure 14: (a) the standard curve of fluorescence as a function of λ-DNA amount ; **(b)** DNA concentration profile for DNA purification. Loading step: S′1; Washing step: S′2~S′3; Elution step: S′ 5~S′11.

6.2.3. Purity Detection

The extracted genomic DNA from whole blood samples by the oxidized porous silicon microchip was used as the temple for PCR to evaluate the quality of purified DNA. The extracted DNA by a commercial kite was used as a positive control and the TE buffer without DNA was as a negative control. In addition, all the amplified results were identified by gel electrophoresis separations. The result is shown in (Fig. **15**), a 203-bp fragment of -gapd gene was successfully amplified. This illustrates that the method for extracting DNA by the microchip was effective to eliminate PCR inhibitors in whole blood and the eluted DNA could be successfully used for PCR reaction with comparable efficiency with the positive control.

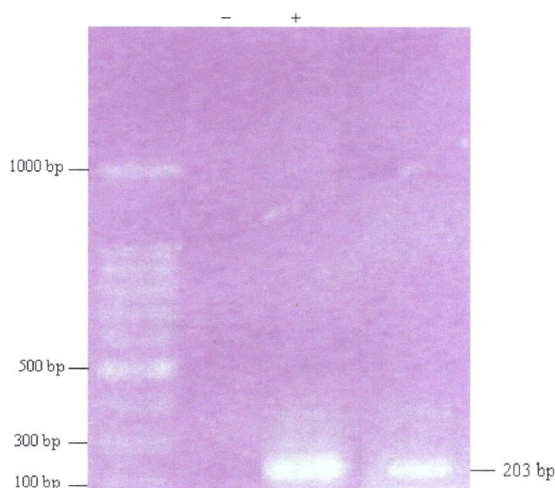

Figure 15: Electrophoresis pattern of PCR products, with both negative and positive controls included. The temple DNA was extracted from rat peripheral blood using the porous matrix. 5μL DNA eluted solution by a commercial kit was used for the positive control reaction.

6.3. DNA Extraction by Using the Rectangle Porous Microchips

6.3.1. DNA Extraction from Prepurified Genomic DNA

Both of the non-porous and porous microchips with the same tortuous rectangle channel were used to recover DNA from 200ng prepurified genomic DNA under the same experimental conditions. As shown in (Fig. **16**), the performance of porous microchip was quite well, with an average of 83% (11.6%RSD) evaluated from five extractions, which was much higher than that of non-porous ones with an average of 39.2% (8.7%RSD). It is important to note that 200ng DNA overloaded the capacity of the non-porous microchip due to the limitation in surface area. Therefore the binding capacity of the non-porous microchip was approximately 75ng/cm^2 with a total internal surface area of 1cm^2. This was essentially in agreement with the results of Cady [15] who found that the binding capacity of SiO$_2$ micropillars was approximately 82ng/cm^2. The previous researches [15, 17] also asserted that the performance of DNA purification chips was determined by the surface areas of the matrices and the extracted efficiency of DNA increased linearly with enhancing surface area. Therefore, compared to the non-porous channel matrix, the binding capacity of the porous channel matrix should be thousands of times higher. However, the practical extraction efficiency was not improved as that much. The reason could probably be that most of the internal pores of the porous matrix might not be used to adsorb DNA and the DNA adsorbed in these pores cannot be easily eluted. Due to the porous matrix directly generated on the internal walls of the channel, the porous microchip can avoid the problem of clogging and high backpressure, and the quantity of the DNA recovery is high enough for PCR and other enzymatic reactions.

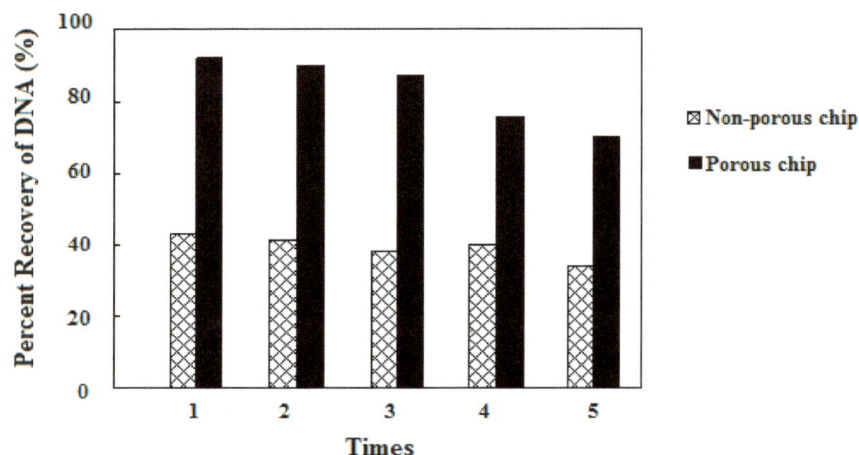

Figure 16: Effect of non-porous and porous microfluidic chips on the efficiency of DNA extraction.

The obtained results demonstrated high recovery and good repeatability of porous microchips for prepurified genomic. However the quick elution of DNA from complex biologic samples was proved to be more challenging. A minimal DNA amount must be guaranteed for PCR amplification and must be confined in a certain volume suitable for microchip-based PCR. Previous researches [1] on microfluidic chips for extracting DNA produced PCR-amplifiable DNA from various samples, and about 80% of the eluted DNA was collected within 10μL. And also Breadmore [13] investigated the influence of load pH and flow rates on the DNA extraction. But quite few reports have focused on the effect of elution process on DNA purification.

Experimental results show that when the surface area of the porous matrix was increased, the difficulty to elute DNA from this matrix also increased correspondingly compared to the non-porous one. One method to improve the elute efficiency was to incubate the DNA absorbed on the porous matrix with the elution buffer for a longer period, and this must be done in a manner that does not increase DNA purification time and therefore it requires the increase of the flow rate in other steps. Since the initial adsorption of DNA on the porous matrix was quite important for the DNA adsorption efficiency, the flow rate of the load step could not increase too much and would be no higher than 4μL/min [12]. However, the flow rate of the washing step could be increased to 15-20μL/min which compensated for the time used in the incubation process. Thus the whole extraction time was about 15min.

6.3.2. DNA Extraction from Rat Peripheral Blood

To determine the effect of incubating temperature on DNA purification efficiency, genomic DNA was extracted from whole rat blood at different incubating temperatures and then examined by fluorescence detection. 10μL of thawed rat blood was mixed with a 100μL loading buffer containing 1% Triton X-100, and 22μL was pumped through the porous microchip at a flow rate of 5μL/min. 45μL of the washing buffer was then pumped through the microchip at a flow rate of 20μL/min. And then 5μL of the TE buffer was introduced and incubated in the microchip for 5min at different temperatures, and followed by the introduction of 25μL of the TE buffer at a flow rate of 10μL/min. During the washing step, fractions were collected every 15μL. During the elution step, fractions were collected every 5μL. Each 5μL collected solution during load, washing and elution steps was mixed with 19μL TE and 1μL SYBR Green I for DNA quantification by fluorescence, respectively. The results of fluorescence detection are shown in (Fig. **17**). The lack of DNA detected in the collected load and washing solutions indicates that DNA molecules were almost completely adsorbed onto the porous matrix. However when the elution temperature changed, the concentration of the eluted DNA varied correspondingly. From (Fig. **17**), the DNA elution efficiency increased first with an increasing temperature. But after a critical temperature, further increase in temperatures induced the decline of DNA elution efficiency. The maximum temperature is around 55°C, which was almost 4 times higher than that of 25°C.

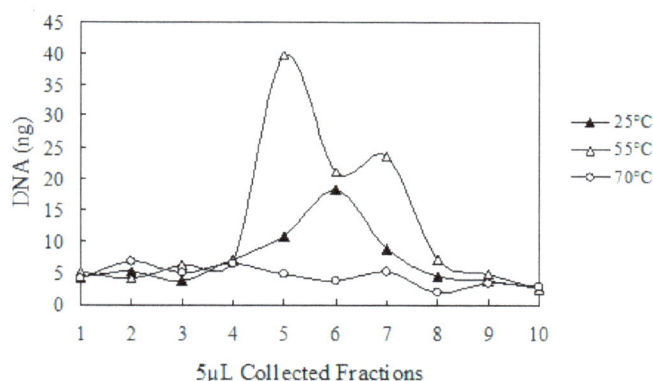

Figure 17: Effect of the elution temperature on DNA purification efficiency. Fraction 1, load step; fractions 2-4, washing step; fraction 5-10, elution step, wherein fraction 5 showed the DNA eluted after incubation for 5min and fractions 6-10 showed the DNA eluted after continuous flowing of the elution buffer.

When the elution temperature was around 55°C , most of DNA (40%) was eluted in the first 5μL TE buffer, consistent with previous results reported by Tian [1]. 49.5ng DNA was purified from 1μL whole rat blood by using the microchip with the porous rectangle channel at the optimal conditions, which was almost twice higher than that by using a commercial kit with silica resin which can only extract about 20-30ng DNA per microlitre blood. The DNA extraction efficiency at 25°C was low due to the inefficient elution of adsorbed DNA molecules from the porous microchannel. The DNA extraction efficiency fell sharply at the incubation temperature of 70°C, since the high temperature might change the DNA structures by disrupting the hydrogen bonding in the double-stranded DNA to produce the single-stranded DNA which could not be detected using SYBR Green I dye.

6.3.3. Purity Detection

Blood is a complex mixture of cells, proteins, peptides, lipids, carbohydrates, and other low molecular weight compounds that are known to inhibit the amplification of DNA by PCR. The extracted genomic DNA from the whole blood at 55°C was compatible with PCR, which further verified that no inhibitory compounds were present. A 203-bp fragment of *-gapd gene* was successfully amplified, which was identified by gel electrophoresis separation. This indicates that the DNA extracted by using the porous microchip had a high purification level and was qualified for subsequent enzymatic reactions. The PCR inhibitors in whole blood samples were removed from the eluted DNA.

Microfluidic chips with porous rectangle channels were fabricated by MEMS and anodization technologies, which were successfully used for purifying genomic DNA from the whole blood. The optimal anodization conditions of the porous channels with complete oxidization were as follows: HF concentration of 10%, current density of 20 mA/cm^2 and etching time of 15 min. Though both porous V-type and rectangular channels were prepared, the latter showed a more uniform morphology and higher stable surface microstructures. Under these optimal conditions, the porous rectangle channels had a surface area of 400 m^2/g and the pore sizes were about 20–30 nm. 83% DNA was recovered by using this porous microchip, which is much higher than 39.2% DNA extraction rate by the non-porous chip. 49.5 ng genomic DNA was purified from one microlitre of whole blood by using the porous microchip at the optimal incubation temperature of 55°C within 15 min, which was twice higher than commercial kits. The integration of the DNA purification chip with other microdevices, such as the hybridization chips, PCR chips and CE chips, could enable the transition of these devices from labs to real applications in the field of point-of-care.

6.4 DNA Extraction from Rat Mesenchymal Stem Cells (MSC)

Microfluidic chips with non-porous matrix (Fig. **18a**), porous matrix (Fig. **18b**) and micropillar based matrix (Fig. **18c**) were used to extract genomic DNA from the cultured rat Mesenchymal Stem Cells (MSC) under developed optimal experimental conditions.

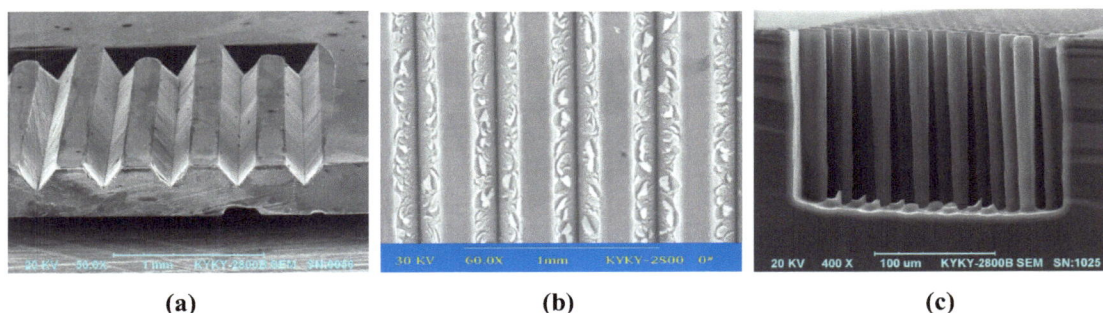

(a)   (b)   (c)

Figure 18: SEM micrographs of v-type microchannels without porous matrix **(a)**, v-type microchannels with porous matrix **(b)** and microchannels with micropillar arrays **(c)**.

30μL of culture solution with about 10^6 cells was mixed with a 150μL loading buffer first. And then 30μL of the mixture solution was pumped through these three types of microfluidic chips. 60μL of the washing buffer was pumped through the chips to remove other inhibitory molecules, followed by the 5 min introduction of 5μL of the TE buffer at 55°C. In the end, 25μL of the TE buffer was pumped through the microchips continuously for DNA elution. During the load step, cells were lyzed by using the lysis reagent, namely Triton X-100. Then genomic DNA was released and DNA, proteins and possible PCR inhibitors were adsorbed onto the matrix. During the washing step, DNA was still adsorbed on the matrix, while proteins and possible PCR inhibitors were removed by loading the wash buffer through the microchannels. During the elution step, the purified DNA was desorbed from the matrix to the TE buffer.

In this experiment, the genomic DNA was extracted in less than 20min with all the collected fractions under fluorescence detection.

As shown in (Fig. **19**), the performance of the microchip with porous matrix was quite well, with an average of 193.7ng DNA extracted, evaluated with data from three extraction samples, and the eluted DNA amount is much higher than that of the non-porous chip with 60.8ng (RSD11.4%) DNA extraction. The performance of the micropillar chip was unsteady and the chip was completely clogged in the third run of

DNA extraction, with 198.9ng DNA collected in the first extraction. According to the experiments, it is worth noting that the DNA of 10^6 cells overloaded the non-porous chip for limitation of the surface area. The surface area of the non-porous chip was about $0.87cm^2$, with the maximal binding capacity of roughly $69.9ng/cm^2$. This is verified by the results of Cady [15] who found that the binding capacity of non-porous micropillars was approximately $82ng/cm^2$.

Previous researches suggested that the performance of the DNA extraction chip was determined by the surface area of the matrix and the extracted DNA was found to increase linearly with the surface area [15, 17]. The binding capacity of the porous matrix would be hundreds or thousands of times increasing, however the extraction efficiency was not improved that much. The potential reasons could be that most of the internal pores and smaller pores might not be used to adsorb DNA and the DNA adsorbed in these pores might not be easily eluted.

The extraction of genomic DNA from a crude biological sample must be PCR-amplifiable. The lysing cells are a complex mixture of proteins, peptides, lipids, carbohydrates, and other low molecular weight compounds that are known to inhibit DNA amplification by PCR. The extracted genomic DNA from rat Mesenchymal Stem Cells (MSC) by using the porous matrix was tested for PCR to ensure that no inhibitory compounds were present. A 203-bp -*gapd* gene of rat cultured cells was successfully amplified.

6.5. DNA Extraction by Using KI as the Binding Salt

Guanidine was used as the binding salt for extracting DNA with these SPE microchips. However, it was proved to be an inhibitor of Polymerase Chain Reaction (PCR) [15]. What is more, guanidine is toxic and it might contaminate the surrounding environment, and it is necessary to look for a new binding salt.

In this section, we reported a high purification DNA extraction method with a SPE microfluidic chip using KI as the new binding salt.

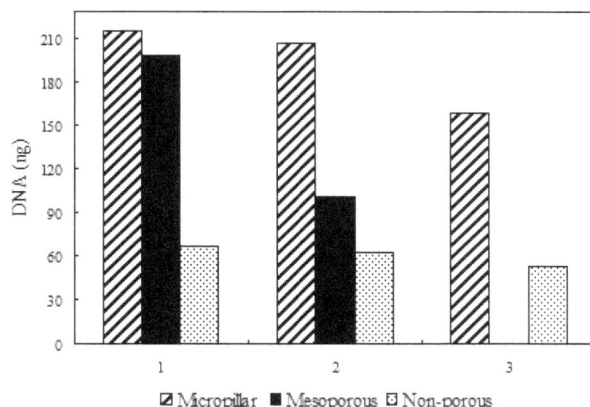

Figure 19: Effect of microfluidic chips with non-porous matrix, mesoporous matrix and micropillar array matrix for extracting DNA from cultured rat Medulla Stem Cells (MSC).

KI was used to replace guanidium for extracting high purity DNA. To evaluate the effect of DNA purification using KI in microchips, 4 mol/L of GuSCN and 5 mol/L of KI were used to bind DNA from whole blood with our SPE microfluidic chip. The concentrations of both binding salts were investigated for the realization of the optimal binding. The extracted DNA was detected by a fluorescence array for amount quantification and also used as the template in PCR. The products of PCR were run through gel electrophoresis for purity evaluation. As seen in (Fig. **20a**), 2 μL whole blood was mixed in 20μL KI or GuSCN binding buffers, respectively, then loaded into the chip (fraction 1, 2), washed with ethanol (fractions 3-4), and eluted with the TE buffer (fractions 5-10). About 13.9 ng genomic DNA was extracted from 1 μL rat whole blood by KI.

A 203-bp fragment of Gapd gene was amplified, illustrating that the eluted genomic DNA from blood was PCR-amplifiable. It was noticed that the purity of the extracted DNA by KI was higher than that by GuSCN (Fig. **20b**). The potential reasons was that the guanidine binding salt, compared to KI, is a stronger denaturant of protein and may behave as an inhibitor of Taq polymerase. Moreover the residual GuSCN in the eluted solution may alter the pH by affecting the dissociation of the PCR solution. In addition the amplification efficiency of PCR for DNA extraction by KI was comparable to that of commercial kits (Fig. **20b**).

Using the conventional centrifugation technology, the SPE method for extracting DNA could effectively remove various buffers and most contaminations. But in a microchip, DNA was extracted by the fluxion of solutions, which means that all the buffers and samples, including some PCR inhibitors, such as guanidine, could not be removed effectively during the flowing process, and a small quantity of them would remain in the eluted solution. Although GuSCN binding salt could extract DNA with a higher recovery amount, the purity of the extracted DNA might be compromised by the presence of the residual GuSCN. At the same time, using KI instead of guanidine could promote DNA binding to silica, which offers another advantage that KI is a nontoxic, safe and cheap salt.

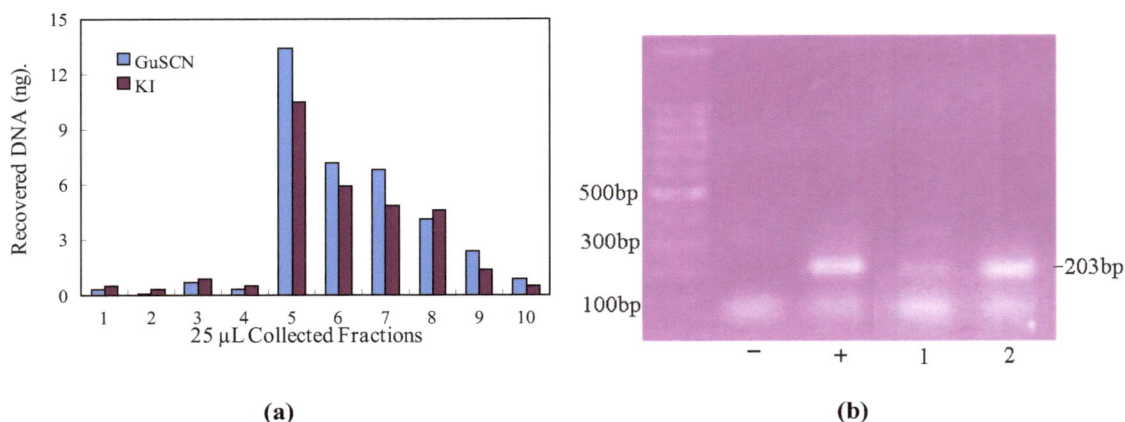

(a)   (b)

Figure 20: (a) DNA elution profile for genomic DNA from rat whole blood. (KI: 5 mol/L KI in TE solution, pH 6.4; GuSCN: 4 mol/L GuSCN in TE solution, pH 6.4). **(b)** Electrophoresis pattern of PCR products, with both negative and positive controls included. The temple DNA was extracted from rat peripheral blood by using a porous microchip. 1 and 2 showed the result of PCR for DNA extraction using GuSCN and KI, respectively. Purified DNA using a commercial kit was used for the positive control.

7. CONCLUSION

From above researches, microfluidic chips with a porous solid phase matrix fabricated by MEMS and anodization technologies were able to extract enough polymerase chain reaction-amplifiable DNA from the whole blood or other biological samples. This highly efficient, effortless, and flexible technology can be used as a key component for initial biologic sample preparation, which will lay the foundation for the realization of μTAS.

ACKNOWLEDGEMENT

The authors greatly acknowledge the financial support from the National Science Foundation of China under Grant numbers 60701019 and 60501020.

REFERENCES

[1]   Tian H, Hühmer AFR, Landers JP. Evaluation of silica resins for direct and efficient extraction of dna from complex biological matrices in a miniaturized format. Anal Biochem 2000; 283(2): 175-91.

[2]   Kutter J, Jacobson S, Ramsey J. Solid phase extraction on microfluidic devices. J Microcolumn 2000; 12(2): 93-7.

[3]   Ekström S, Malmstr m J, Wallman L, *et al*. On-chip microextraction for proteomic sample preparation of in-gel digests. Proteomics 2002; 2(4): 413-21.

[4]   Bergkvist J, Ekstr m S, Wallman L, *et al*. Improved chip design for integrated solid-phase microextraction in on-line proteomic sample preparation. Proteomics Clin Appl 2002; 2(4): 422-9.

[5]   Yu C, Davey M, Svec F, Frechet J. Monolithic porous polymer for on-chip solid-phase extraction and preconcentration prepared by photoinitiated *in situ* polymerization within a microfluidic device. Anal Chem 2001; 73(21): 5088-96.

[6]   Wen J, Legendre LA, Bienvenue JM, Landers JP. Purification of nucleic acids in microfluidic devices. Anal Chem 2008; 80(17): 6472-9.

[7]   Boom R, Sol C, Salimans M, *et al*. Rapid and simple method for purification of nucleic acids. J Clin Microbiol 1990;28(3):495-503.

[8]   Vogelstein B, Gillespie D. Preparative and analytical purification of DNA from agarose. Proc Nat Acad Sci 1979; 76(2): 615-9.

[9]   Marko M, Chipperfield R, Birnboim H. A procedure for the large-scale isolation of highly purified plasmid DNA using alkaline extraction and binding to glass powder. Anal Biochem 1982; 121: 382-7.

[10]   Wen J, Guillo C, Ferrance JP, Landers JP. DNA extraction using a tetramethyl orthosilicate-grafted photopolymerized monolithic solid phase. Anal Chem 2006; 78(5): 1673-81.

[11]   Wen J, Guillo C, Ferrance JP, Landers JP. Microfluidic-based DNA purification in a two-stage, dual-phase microchip containing a reversed-phase and a photopolymerized monolith. Anal Chem 2007; 79(16): 6135-42.

[12]   Wolfe KA, Breadmore MC, Ferrance JP, *et al*. Toward a microchip-based solid-phase extraction method for isolation of nucleic acids. Electrophoresis 2002; 23(5): 727-33.

[13]   Breadmore MC, Wolfe KA, Arcibal IG, *et al*. Microchip-based purification of DNA from biological samples. Anal Chem 2003; 75(8): 1880-6.

[14]   Wu Q, Bienvenue JM, Hassan BJ, *et al*. Microchip-based macroporous silica sol-gel monolith for efficient isolation of DNA from clinical samples. Anal Chem 2006; 78(16): 5704-10.

[15]   Cady NC, Stelick S, Batt CA. Nucleic acid purification using microfabricated silicon structures. Biosens Bioelectron 2003; 19(1): 59-66.

[16]   Poeckh T, Lopez S, Fuller A, Solomon M, Larson R. Adsorption and elution characteristics of nucleic acids on silica surfaces and their use in designing a miniaturized purification unit. Anal Biochem 2008; 373(2): 253-62.

[17]   Christel LA, Petersen K, McMillan W, Northrup MA. Rapid, automated nuleic acid probe assays using silicon microstructures for nucleic acid concentration. J Biomech Eng 1999; 121(1): 22-7.

[18]   Cady N, Stelick S, Kunnavakkam M, Batt C. Real-time PCR detection of Listeria monocytogenes using an integrated microfluidics platform. Sensor Actuator B Chem 2005; 107(1): 332-41.

[19]   Chen X, Cui DF, Liu CC. On-line cell lysis and DNA extraction on a microfluidic biochip fabricated by microelectromechanical system technology. Electrophoresis 2008; 29(9): 1844-51.

[20]   Chen X, Cui DF, Liu CC, Li H. Microfabrication and characterization of porous channels for DNA purification. J Micromech Microeng 2007; 17(1): 68-75.

[21]   Nakagawa T, Hashimoto R, Maruyama K, *et al*. Capture and release of DNA using aminosilane-modified bacterial magnetic particles for automated detection system of single nucleotide polymorphisms. Biotechnol Bioeng 2006; 94(5): 862-8.

[22]   Cao W, Easley CJ, Ferrance JP, Landers JP. Chitosan as a polymer for pH-induced DNA capture in a totally aqueous system. Anal Chem 2006; 78(20): 7222-8.

[23]   Satterfield BC, Stern S, Caplan MR, Hukari KW, West JAA. Microfluidic purification and preconcentration of mRNA by flow-through polymeric monolith. Anal Chem 2007; 79(16): 6230-5.

[24]   Irimia D, Mindrinos M, Russom A, *et al*. Genome-wide transcriptome analysis of 150 cell samples. Integr Biol 2009; 1: 99-107.

[25]   Xu Y, Vaidya B, Patel AB, *et al*. Solid-phase reversible immobilization in microfluidic chips for the purification of dye-labeled DNA sequencing fragments. Anal Chem 2003; 75(13): 2975-84.

[26]   Elkin C, Kapur H, Smith T, *et al*. Magnetic bead purification of labeled DNA fragments for high-throughput capillary electrophoresis sequencing. Biotechniques 2002; 32(6): 1296-302.

[27]   DeAngelis M, Wang D, Hawkins T. Solid-phase reversible immobilization for the isolation of PCR products. Nucleic Acids Res 1995; 23(22): 4742-3.

[28] Elkin C, Richardson P, Fourcade H, *et al.* High-throughput plasmid purification for capillary sequencing. Genome Res 2001; 11(7): 1269.

[29] Hawkins T, O'Connor-Morin T, Roy A, Santillan C. DNA purification and isolation using a solid-phase. Nucleic Acids Res 1994; 22(21): 4543-4.

[30] Witek MA, Llopis SD, Wheatley A, McCarley RL, Soper SA. Purification and preconcentration of genomic DNA from whole cell lysates using photoactivated polycarbonate (PPC) microfluidic chips. Nucleic Acids Res 2006; 34(10): e74.

[31] Witek MA, Hupert ML, Park DSW, *et al.* 96-Well polycarbonate-based microfluidic titer plate for high-throughput purification of DNA and RNA. Anal Chem 2008; 80(9): 3483-91.

[32] Kessler H, Muhlbauer G, Stelzl E, *et al.* Fully automated nucleic acid extraction: MagNA Pure LC. Clin Chem 2001; 47(6): 1124-6.

[33] Akutsu J, Tojo Y, Segawa O, *et al.* Development of an integrated automation system with a magnetic bead-mediated nucleic acid purification device for genetic analysis and gene manipulation. Biotechnol Bioeng 2004; 86(6): 667-71.

[34] Zaytseva N, Montagna R, Baeumner A. Microfluidic biosensor for the serotype-specific detection of Dengue virus RNA. Anal Chem 2005; 77(23): 7520-7.

[35] Yeung SW, Hsing IM. Manipulation and extraction of genomic DNA from cell lysate by functionalized magnetic particles for lab on a chip applications. Biosens Bioelectron 2006; 21(7): 989-97.

[36] Lien K, Liu C, Kuo P, Lee G. Microfluidic system for detection of α-thalassemia-1 deletion using saliva samples. Anal Chem 2009; 81(11): 4502-9.

[37] Chen Z, Mauk MG, Wang J, *et al.* A Microfluidic System for Saliva-Based Detection of Infectious Diseases. Ann N Y Acad Sci 2007; 1098(1): 429-36.

[38] Jiang G, Harrison DJ. mRNA isolation in a microfluidic device for eventual integration of cDNA library construction. Analyst 2000; 125(12): 2176-9.

[39] Zaytseva N, Goral V, Montagna R, Baeumner A. Development of a microfluidic biosensor module for pathogen detection. Lab Chip 2005; 5(8): 805-11.

[40] Tokeshi M, Minagawa T, Kitamori T. Integration of a microextraction system on a glass chip: ion-pair solvent extraction of Fe (II) with 4, 7-diphenyl-1, 10-phenanthrolinedisulfonic acid and tri-n-octylmethylammonium chloride. Anal Chem 2000; 72(7): 1711-4.

[41] Hisamoto H, Horiuchi T, Uchiyama K, *et al.* On-chip integration of sequential ion-sensing system based on intermittent reagent pumping and formation of two-layer flow. Anal Chem 2001; 73(22): 5551-6.

[42] Reddy V, Zahn JD. Interfacial stabilization of organic-aqueous two-phase microflows for a miniaturized DNA extraction module. J Colloid Interface Sci 2005; 286(1): 158-65.

[43] Melzak KA, Sherwood CS, Turner RFB, Haynes CA. Driving forces for DNA adsorption to silica in perchlorate solutions. J Colloid Interface Sci 1996; 181(2): 635-44.

[44] Uhlir A. Electrolytic shaping of germanium and silicon. Bell Syst Tech J 1956;35(2):333-47.

[45] Turner DR. Electropolishing silicon in hydrofluoric acid solutions. J Electrochem Soc 1958; 105: 402-8.

[46] Bisi O, Ossicini S, Pavesi L. Porous silicon: A quantum sponge structure for silicon based optoelectronics. Surf Sci Rep 2000; 38(1): 1-126.

[47] Schmuki P, Schlierf U, Herrmann T, Champion G. Pore initiation and growth on n-InP (100). Electrochim Acta 2003; 48(9): 1301-8.

[48] Bilyalov R, Stalmans L, Beaucarne G, *et al.* Porous silicon as an intermediate layer for thin-film solar cell. Sol Energy Mater Sol Cells 2001; 65(1-4): 477-85.

[49] Menna P, Di Francia G, La Ferrara V. Porous silicon in solar cells: a review and a description of its application as an AR coating. Sol Energy Mater Sol Cells 1995; 37(1): 13-24.

[50] Stalmans L, Poortmans J, Bender H, *et al.* Porous silicon in crystalline silicon solar cells: a review and the effect on the internal quantum efficiency. Progress in Photovoltaics Research and Applications 1998; 6(4): 233-46.

[51] Park J, Lee J. Characterization of 10μm thick porous silicon dioxide obtained by complex oxidation process for RF application. Materials Chemistry & Physics 2003; 82(1): 134-9.

[52] Lopez H, Fauchet P. Erbium emission from porous silicon one-dimensional photonic band gap structures. Appl Phys Lett 2000; 77: 3704-6.

[53] Massera E, Nasti I, Quercia L, Rea I, Di Francia G. Improvement of stability and recovery time in porous-silicon-based NO2 sensor. Sensor Actuator B Chem 2004; 102(2): 195-7.

[54] Björkqvist M, Salonen J, Paski J, Laine E. Characterization of thermally carbonized porous silicon humidity sensor. Sens Actuators A 2004; 112(2-3): 244-7.

[55] Lin V, Motesharei K, Dancil K, Sailor M, Ghadiri M. A porous silicon-based optical interferometric biosensor. Science 1997; 278(5339): 840-3.

[56] Melander C, Bengtsson M, Schagerl f H, *et al.* Investigation of micro-immobilised enzyme reactors containing endoglucanases for efficient hydrolysis of cellodextrins and cellulose derivatives. Anal Chim Acta 2005; 550(1-2): 182-90.

[57] Bengtsson M, Ekström S, Marko-Varga G, Laurell T. Improved performance in silicon enzyme microreactors obtained by homogeneous porous silicon carrier matrix. Talanta 2002; 56(2): 341-53.

[58] Clicq D, Tjerkstra R, Gardeniers J, *et al.* Porous silicon as a stationary phase for shear-driven chromatography. J Chromatogr A 2004; 1032(1-2): 185-91.

[59] Steinhauer C, Ressine A, Marko-Varga G, *et al.* Biocompatibility of surfaces for antibody microarrays: design of macroporous silicon substrates. Anal Biochem 2005; 341(2): 204-13.

[60] Chen X, Cui DF, Liu CC, Li H, Zhao WX. Sample pretreatment microfluidic chip for DNA extraction from rat peripheral blood. Chem J Chin U 2006; 27(4): 618-21.

[61] Francia G, Ferrara V, Manzo S, Chiavarini S. Towards a label-free optical porous silicon DNA sensor. Biosens Bioelectron 2005; 21(4): 661-5.

[62] Chan S, Li Y, Rothberg L, Miller B, Fauchet P. Nanoscale silicon microcavities for biosensing. Mater Sci Eng C 2001;15(1-2):277-82.

[63] Ryel Kwon D, Ghosh S, Lee C. Growth and nucleation of pores in n-type porous silicon and related photoluminescence. Mater Sci Eng B 2003; 103(1): 1-8.

[64] Ressine A, Ekstrom S, Marko-Varga G, Laurell T. Macro-/nanoporous silicon as a support for high-performance protein microarrays. Anal Chem 2003; 75(24): 6968-74.

[65] Kan P, Finstad T. Oxidation of macroporous silicon for thick thermal insulation. Mater Sci Eng B 2005; 118(1-3): 289-92.

[66] Cullis A, Canham L, Calcott P. The structural and luminescence properties of porous silicon. J Appl Phys 1997; 82: 909-65.

[67] Dai Z, Xu X, Ju H. Direct electrochemistry and electrocatalysis of myoglobin immobilized on a hexagonal mesoporous silica matrix. Anal Biochem 2004; 332(1): 23-31.

[68] Salonen J, Laitinen L, Kaukonen A, *et al.* Mesoporous silicon microparticles for oral drug delivery: Loading and release of five model drugs. J Control Release 2005; 108(2-3): 362-74.

[69] Lendl B, Schindler R, Frank J, *et al.* Fourier transform infrared detection in miniaturized total analysis systems for sucrose analysis. Anal Chem 1997;69(15):2877-81.

[70] Drott J, Rosengren L, Lindström K, Laurell T. Porous silicon carrier matrices in micro enzyme reactors-influence of matrix depth. Microchimica Acta 1999; 131(1): 115-20.

[71] Barthlott W, Neinhuis C. Purity of the sacred lotus, or escape from contamination in biological surfaces. Planta 1997; 202(1): 1-8.

[72] Chen X, Cui D, Liu C, Cai H. Fabrication of solid phase extraction DNA chips based on bio-micro-electron-mechanical system technology. Chin J Anal Chem 2006. p. 433-7.

[73] Chen X, Cui DF, Liu CC. High purity DNA extraction with a SPE microfluidic chip using KI as the binding salt. Chin Chem Lett 2006; 17(8): 1101-4.

[74] Chen X, Cui DF, Liu CC, Li H. Fabrication of DNA purification microchip integrated with mesoporous matrix based on MEMS technology. Microsys Technol 2008; 14(1): 51-7.

[75] Torchinskaya T, Korsunskaya N, Khomenkova L, Dhumaev B, Prokes S. The role of oxidation on porous silicon photoluminescence and its excitation. Thin Solid Films 2001; 381(1): 88-93.

CHAPTER 6

Integrated Microfluidic Chips for Whole Blood Pretreatment

Xing Chen*, Dafu Cui and Jian Chen

State Key Laboratory of Transducer Technology, Institute of Electronics, Chinese Academy of Sciences, Beijing, China

Abstract: Integrated on-chip whole blood sample pretreatment includes blood cell separation, cell lysis and DNA extraction. Compared to macro-counterparts, whole blood sample pretreatment using integrated microfluidic devices has the advantages of low sample consumption, short time-scale measurement and high portability. In this chapter, reported integrated microfluidic devices for blood sample preparation are briefly summarized and compared. And then a specific microfluidic chip with the function of cell separation, cell lysis and DNA extraction is proposed for detailed discussion. The integrated microfluidic device is capable of pre-treating whole blood samples, paving ways for the eventual integration of sample preparation, PCR, and electrophoresis on a single chip in the field of point-of-care genetic analysis, environmental testing, and biological warfare agent detection.

Keywords: Integration, MEMS, microfluidics

1. INTRODUCTION

Micro Total Analysis System (μTAS), also named as Lab-on-a-Chip, uses MEMS technologies to fabricate micro biochemical analyzing systems on the surface of solid chips [1-9]. It can detect and measure large amounts of inorganic ions, organic matters, proteins, nucleic acids, and other biochemical ingredients quickly and precisely.

Theoretically μTAS is a micro-biochemical analyzing instrument for the integration of three key components: sample preprocessing, biochemical analyzing and relevant chemical reactions, and detection of results on a single microchip. It has many advantages such as low cost, high throughput, high speed, analyzing process automation, low reagent consumption and easy integration. It can be widely used in biomedicine, remedy filtration, food sanitation, environment inspection and many other areas, which takes the lead in the future trends of analyzing instruments: minimization and integration.

However, up to now, integration is the bottleneck in the development of μTAS, which is also one of the research focuses in the area of microfluidics. Although the design methodology of chip integration is quite diverse, the requirements are the same which are steady flow in the chip, minimized dead volume, nonvolatile liquid, no bubble formation, consistent material properties, no interfere signal for detection, high signal-to-noise ratio and proper dispose of the wasted fluid and chips. Previously published integration methods can be classified into two categories, which are blend integration and whole chip integration, respectively.

In blend integration separated parts are used to construct μTAS. This integration method is bottom-up, connecting various separated chips of multiple functions on the board of a micro platform. On the other hand, whole chip integration uses the top-down design method by integrating the sample pretreatment, analysis and detection together on one single chip. Compared to whole chip integration, blend integration doesn't need the integration of functional components on a single chip and has the advantages of low cost and high quality. However, blend integration is not truly "compatible" with the concept of "lab on chip" since it needs off-chip detection systems such as optical detection instruments, and therefore is confined in the lab. In the meanwhile, whole chip integration features on-chip sample pretreatment, analysis and detection, with the potential of forming micro total analysis systems.

---

*Address correspondence to Xing Chen; State Key Laboratory of Transducer Technology, Institute of Electronics, Chinese Academy of Sciences, Beijing 100190, China; phone and fax: +86-10-58887188; E-mail: xchen@mail.ie.ac.cn

Currently several microfluidic chips have already shown the signs of integration. For example, PCR amplification and capillary electrophoresis have been implemented in one microfluidic chip [10-14]. Also, on-chip detection components have been integrated into the microfluidic platforms with other functional components [11, 15, 16].

However, μTAS is not "micro" enough, and far from "total" with very low integration level. One bottleneck in the development of μTAS is to integrate on-chip sample preprocessing components such as whole blood sample pretreatment with the analysis and detection units. In this chapter, reported integrated microfluidic devices for whole blood pretreatment are briefly introduced and compared. And then a case study on an integrated microfluidic device is put forward, where all the key pretreatment steps are conducted in the continuous flow mode.

2. CURRENT INTEGRATED MICROFLUIDIC PLATFORMS FOR WHOLE BLOOD PRETREATMENT

The first integrated silicon microchip with discrete components was developed and demonstrated for the extraction of the human genomic deoxyribonucleic acid (DNA) from blood samples [17]. The silicon chip consists of a micromixer, a microfilter, a microbinder and two microvalves to perform cell separation, cell lysing to DNA extraction subsequently, shown in (Fig. **1**). Raw blood samples, including White Blood Cells (WBCs), Red Blood Cells (RBCs), platelets, plasma, *etc.* are pumped directly into the microchip for sample pretreatment. For DNA extraction, only WBCs with the DNA are isolated or captured based on a size-exclusion principle using a microfabricated pillar structure filter. The spiral microchannel is used for solution mixing to lyse cells chemically and extract DNA macromolecules. With the high salt concentration in the lysis buffer, the DNA molecules selectively bind to the microbinder's surface based on the principle of Solid Phase Extraction (SPE). Experimental results showed that this chip was capable of collecting about 10 ng of gDNA from 1 μL of human blood.

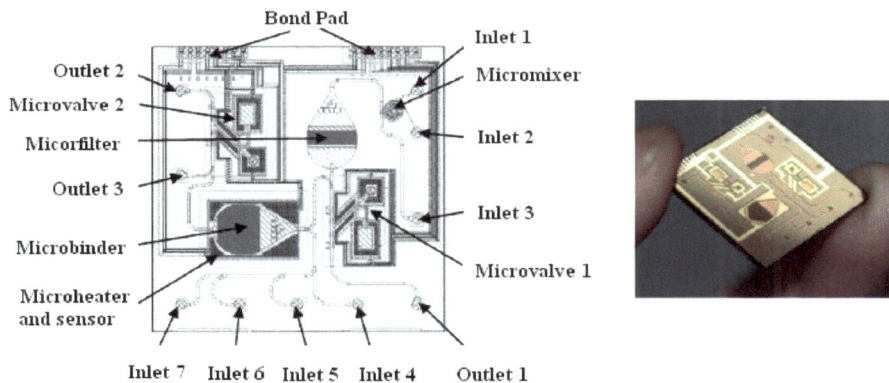

Figure 1: The integrated DNA extraction chip with the dimension of 2 cm×2 cm [17]. Reprinted with permission from Elsevier. Copyright (2007)

Another microfluidic chip for automated, parallel nucleic acid extraction from bacteria or mammalian cells was developed with 26 access holes and 54 valves [18]. All the functions including cell sorting, cell lysis and DNA purification are included on the single microfluidic chip. The chip was fabricated by multilayer soft lithography while in the design, modified magnetic beads are used to isolate cells and extract nucleic acid molecules. All the pretreatment processes are controlled by micromechanical valves.

In another example, by integrating a mechanical/chemical cell lysis system, genomic DNA was purified from gram positive and gram negative bacteria using microscale silica bead/polymer composites [19]. In this device, bacteria were lysed on-chip *via* a hybrid chemical/mechanical method. Once lysed, the bacterial DNA was isolated using a microscale silica bead/polymer composite solid-phaseextraction (SPE) column. Approximately 50 ng/mL DNA was extracted from 10^8 *E. faecalis*, 10^8 *E. coli* and from 10^7 *B. subtilis*.

Furthermore, a fully integrated microfluidic genetic analysis system was also developed for PCR and electrophoretic detection, used to locate the presence of infectious pathogens [20].

3. CASE STUDY

3.1. Principle and Method

In this section, a specific example on an integrated sample pre-treatment microfluidic device is put forward to further demonstrate the concept of integrated microfluidic chips for whole blood sample pretreatment. In this microfluidic chip, a silicon wafer, a PDMS-glass compound and a PMMA plate are used for device construction through a simple and reliable fabrication process. The integrated microfluidic device consists of a microfilter, a micromixer, a micropillar array, a microweir, and porous matrix targeting sample pretreatment of whole blood. Cell separation, cell lysis and DNA purification are performed in this miniaturized device during a continuous flow process.

In this example, crossflow filtration is proposed to separate blood cells, which successfully addressed the issue of clogging or jamming [21-24]. After blood cells are chemically lyzed in guanidine buffer, genomic DNA molecules in WBCs are released and adsorbed on porous matrix fabricated by anodizing silicon in the HF/ethanol electrolyte [25]. Utilizing the feature of continuous flow, the proposed microfluidic device enables the cell separation, white blood cell lysis and DNA purification with high speed [26].

3.2. Design and Fabrication

(Fig. **2**) presents a schematic view of the integrated microfluidic device, which consists of two major modules, namely a crossflow microfilter and a component for cell lysis and DNA purification. In the module of cell separation, WBCs are separated from whole blood samples based on the crossflow filtration and then transported to the module of cell lysis and DNA purification, while RBCs and blood plasma are removed. In the module of cell lysis, the purified WBCs are lyzed chemically, and genomic DNA molecules are purified by means of solid phase extraction.

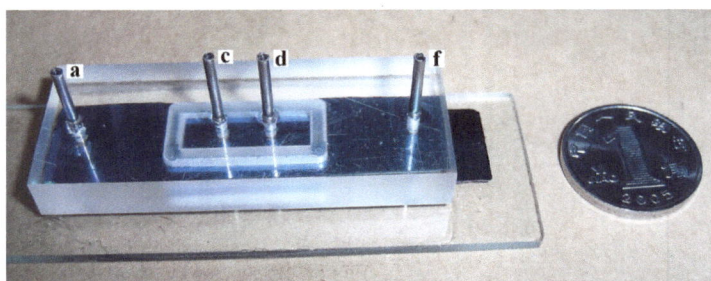

Figure 2: Photograph of the integrated microfluidic device.

In the module of cell separation, WBCs and RBCs are separated by means of the crossflow filtration, based on cell size difference. Compared with dead-end filtration, crossflow filtration avoids the problems of clogging or jamming. It is worth noting that the removed RBCs and blood plasma from WBCs could be further used in other biochemical analysis in the future. In the cell lysis and DNA purification module, the separated WBCs are lyzed in a continuously flowing lysis buffer, and after cell lysis, their DNA molecules are absorbed on the surface of microchannels. This module is designed to sandwich WBCs in lysis buffer to ensure that each cell is exposed to the lysis buffer for efficient cell lysis. Moreover the porous silicon layer fabricated from the internal surface of the microchannel is used to increase the surface area and therefore enhance the efficiency of solid phase extraction.

As shown in (Fig. **3**), the integrated microfluidic device is made of tree layers. The top layer is made of PMMA with a WBC reservoir, two input ports and two output ports. The middle layer is a silicon plate with microstructures for cell separation, cell lysis and DNA purification. The bottom layer is a glass-PDMS compounded cover for forming closed microchannels.

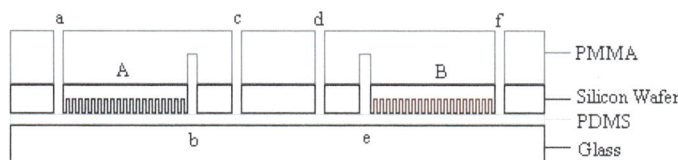

Figure 3: Schematic view of the integrated microfluidic device.

The fabrication process of the compound cover is shown as follows. A flat glass substrate was exposed to a vapor of trimethylchlorosilane for 5min in order to facilitate the release of the glass substrate. A 10:1 mixture of PDMS oligomer and cross-linking agent, which had been degassed under vacuum, was poured onto the glass substrate. Then another glass was put on the PDMS prepolymer carefully and cured in an oven at 80°C for 1h. After the silanized glass was peeled off, the compounded cover was successfully fabricated. To pump blood samples and the lysis buffer into microchips, two peristaltic pumps (Scientific Support, Inc., Hayward, CA) were connected to the input ports, while two centrifugal tubes were connected to the output ports for collecting RBCs and DNA purification buffers, respectively.

The silicon wafer layer consists of a section for cell separation, a section for cell lysis and DNA purification, and six holes as inlets and outlets. In the region of cell separation, one tortuous channel 30μm deep, 200μm wide, 160mm long, is divided into three subchannels spaced by two rows of filtration barriers in parallel. Pillar-type filtration barriers are shown in (Fig. **4A**), in which the two rows of pillars are about 20μm in diameter and spaced by 6.5μm. Weir-type filtration barriers are shown in (Fig. **4B**), in which the two weirs are 26.5μm high and 20μm wide. Hence the separation gap is 6.5μm for pillar-type filtration barriers and 3.5μm for weir-type filtration barriers. In the region of cell lysis and DNA purification, the second tortuous channel that is 30μm deep, 200μm wide, 160mm long, is fabricated and coated with a porous layer.

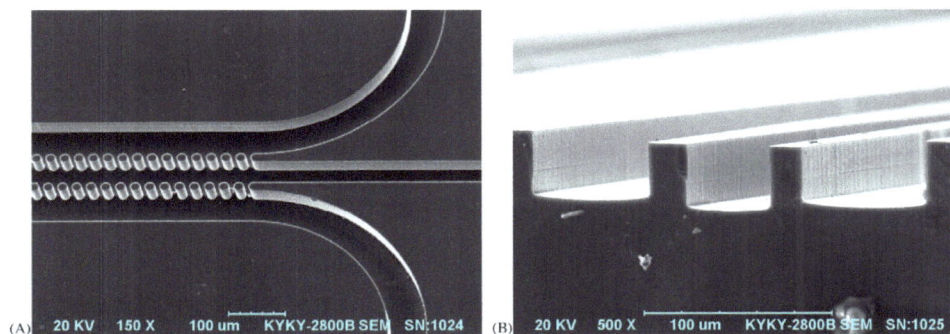

Figure 4: SEM micrographs of filtration microstructures in the region of cell separation **(A)** a micropillar array in the area close to the end and **(B)** cross-sectional view of the weir structure.

The process used to fabricate the silicon plate is as follows. Fabrication started with the n-type double-side polished silicon wafers of (100) crystal orientation, with the thickness of approximately 350μm. The silicon wafers were first coated with 0.3μm thick of silicon nitride, which was used as a mask for patterning holes and fabricating the porous silicon layer on the internal walls of microchannels. The backside was patterned and etched in KOH (33%) solution at 80°C for 5h to produce six fluidic ports that were 300 μm deep. And then the front side was patterned with the DRIE mask, and etched for approximately 5min in an AMS100 SE DRIE (deep reactive ion etching) etcher (Adixen, France) to form micro pillars and channels simultaneously. Weir-type filtration barriers and channels were fabricated separately based on a two-mask process. By means of the first photolithography, weir structures were patterned on the silicon wafer and then microchannels were formed in the second photolithography. Then a surface-enlarging porous silicon layer on the internal walls of microchannels only in the region of cell lysis and DNA purification was produced by the anodisation etching technology. For each silicon substrate, the anodisation was performed in an ethanol (95%) / HF (48%) solution (mixing ratio 2:3) at three current densities of 8, 30, 80 mA/cm^2, respectively. The anodisation process was kept for 15 min. After the process of anodisation etching, the silicon wafers were thoroughly rinsed in distilled water. Then the wafers were put into a furnace where the temperature was increased up 1050°C from room temperature within 2 hours and then kept in the steady high temperature for another hour, to cover porous silicon layer completely with porous silicon dioxide layer. Following the oxidation process, the patterned silicon nitride mask was removed by plasma etching.

3.3. Sample Pretreatment

Three devices with the weir-type filtration structures for cell separation and the porous SiO_2 layer for DNA purification were tested for pre-treating whole blood. 5 µL whole blood was diluted in 0.9% NaCl solution and pumped into each device. WBCs were collected in the WBC port and still confined in the device since they cannot pass through the filtration structures due to size limitations, whereas RBCs were noticed to pass through the filter area and separated from WBCs. After the blood cells were separated, the buffer containing of lysing reagents and bonding salt was pumped into the microdevice and mixed with WBCs stored in the device in the continuous flowing mode. WBCs were gradually lyzed during the continuous flowing process, and the DNA molecules in WBCs were released and adsorbed on the porous matrix under a high ionic strength chaotropic condition. Following cell lysis, the washing buffer and elution buffer were pumped into each device sequentially. In the end, the purified DNA molecules were desorbed with the elution buffer and collected at the outlet port. The whole process of blood pretreatment took less than 50 min. The amount of DNA molecules in all the collected fractions during the elution step was quantified using SYBR Green I in fluorescence detection.

Fluorescence detection shows that 36.7 ng, 31.8 ng, and 38.6 ng genomic DNA were extracted from 1 µL whole blood by using three devices respectively with quite consistent results. The number of DNA molecules extracted by the integrated device were slightly higher than the values using a commercial kit with silica resin which can only extract about 20~30 ng DNA per microlitre blood.

The extraction of genomic DNA from a crude biological sample should be PCR-amplifiable, or they are not quite meaningful in the clinical detection. However, the lyzed cells are a complex mixture of proteins, peptides, lipids, carbohydrates, and other low molecular weight compounds that are known to inhibit DNA amplification by PCR. The extracted genomic DNA from whole blood samples in three separate experiments, respectively, was tested for PCR amplification to verify that no inhibitory compounds were present. Shown in (Fig. **5**), a 203-bp fragment of gene was successfully amplified, which was identified by gel electrophoresis separations. This illustrates that the purification of eluted DNA molecules by using the microfluidic device is guaranteed for subsequent analysis.

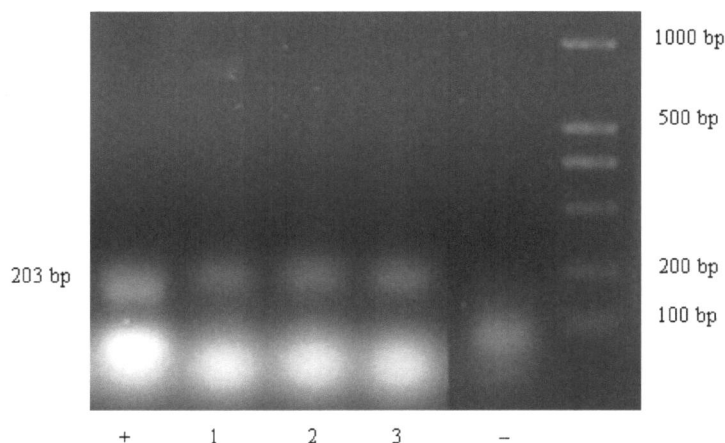

Figure 5: Photograph of electrophoretic gel results. Lanes: (M) molecular weight standard; (+) positive control; (1–3) PCR amplification of collected fractions during the elution step in three separate experiments, respectively; (−) negative.

4. CONCLUSIONS

This example demonstrated a microfluidic device capable of performing cell separation, cell lysis and DNA purification. Using MEMS techniques, this device enables the reduction in sample and reagent consumption, high speed and minimum human intervention during its operation. During continuous flowing process, successful separation of WBCs from RBCs, WBC lysis and genomic DNA purification were demonstrated. In the experiment, no clogging or jamming was noticed. The developed integrated microfluidic device provides a powerful tool for biological sample pre-treatment, and could be used in the near future as an indispensible element of the lab-on-a-chip.

ACKNOWLEDGEMENT

The authors greatly acknowledge the financial support from the National Science Foundation of China under the grant numbers of 60701019 and 60501020.

REFERENCES

[1]   deMello AJ. Control and detection of chemical reactions in microfluidic systems. Nature 2006; 442(7101): 394-402.

[2]   Craighead H. Future lab-on-a-chip technologies for interrogating individual molecules. Nature 2006; 442(7101): 387-93.

[3]   Hogan J. Lab on a chip: A little goes a long way. Nature 2006; 442(7101): 351-2.

[4]   El-Ali J, Sorger PK, Jensen KF. Cells on chips. Nature 2006 J; 442(7101): 403-11.

[5]   Whitesides GM. The origins and the future of microfluidics. Nature 2006; 442(7101): 368-73.

[6]   Janasek D, Franzke J, Manz A. Scaling and the design of miniaturized chemical-analysis systems. Nature 2006; 442(7101): 374-80.

[7]   Psaltis D, Quake SR, Yang CH. Developing optofluidic technology through the fusion of microfluidics and optics. Nature 2006; 442(7101): 381-6.

[8]   Daw R, Finkelstein J. Lab on a chip. Nature 2006;442(7101):367.

[9]   Yager P, Edwards T, Fu E, *et al.* Microfluidic diagnostic technologies for global public health. Nature 2006 ; 442(7101): 412-8.

[10]   Woolley A, Hadley D, Landre P, Mathies R, Northrup M. Functional integration of PCR amplification and capillary electrophoresis in a microfabricated DNA analysis device. Anal Chem 1996; 68(23): 4081-6.

[11]   Lagally E, Simpson P, Mathies R. Monolithic integrated microfluidic DNA amplification and capillary electrophoresis analysis system. Sens Actuators B 2000; 63(3): 138-46.

[12]   Waters L, Jacobson S, Kroutchinina N, *et al.* Microchip device for cell lysis, multiplex PCR amplification, and electrophoretic sizing. Anal Chem 1998; 70(1): 158-62.

[13]   Lagally E, Emrich C, Mathies R. Fully integrated PCR-capillary electrophoresis microsystem for DNA analysis. Lab Chip 2001; 1(2): 102-7.

[14]   Telenius H, Carter N, Bebb C, *et al.* Degenerate oligonucleotide-primed PCR: general amplification of target DNA by a single degenerate primer. Genomics 1992; 13(3): 718.

[15]   Burns M, Johnson B, Brahmasandra S, *et al.* An integrated nanoliter DNA analysis device. Science 1998; 282(5388): 484.

[16]   Blazej R, Kumaresan P, Mathies R. Microfabricated bioprocessor for integrated nanoliter-scale Sanger DNA sequencing. Proc Natl Acad Sci U S A. 2006; 103(19): 7240-5.

[17]   Ji H, Samper V, Chen Y, *et al.* DNA purification silicon chip. Sens Actuators A 2007; 139(1-2): 139-44.

[18]   Hong J, Studer V, Hang G, Anderson W, Quake S. A nanoliter-scale nucleic acid processor with parallel architecture. Nat Biotechnol 2004; 22(4): 435-9.

[19]   Mahalanabis M, Al-Muayad H, Kulinski M, Altman D, Klapperich C. Cell lysis and DNA extraction of gram-positive and gram-negative bacteria from whole blood in a disposable microfluidic chip. Lab Chip 2009; 9: 2811-7.

[20]   Easley CJ, Karlinsey JM, Bienvenue JM, *et al.* A fully integrated microfluidic genetic analysis system with sample-in-answer-out capability. Proc Nat Acad Sci 2006; 103(51): 19272-7.

[21]   Sethu P, Sin A, Toner M. Microfluidic diffusive filter for apheresis (leukapheresis). Lab Chip 2006; 6(1): 83-9.

[22] VanDelinder V, Groisman A. Separation of plasma from whole human blood in a continuous cross-flow in a molded microfluidic device. Anal Chem 2006; 78(11): 3765-71.

[23] Jäggi RD, Sandoz R, Effenhauser CS. Microfluidic depletion of red blood cells from whole blood in high-aspect-ratio microchannels. Microfluidics Nanofluidics 2007; 3(1): 47-53.

[24] Chen X, Cui DF, Liu CC, Li H. Microfluidic chip for blood cell separation and collection based on crossflow filtration. Sens Actuators B 2008; 130(1): 216-21.

[25] Chen X, Cui DF, Liu CC. On-line cell lysis and DNA extraction on a microfluidic biochip fabricated by microelectromechanical system technology. Electrophoresis 2008; 29(9): 1844-51.

[26] Chen X, Cui D, Liu C, Li H, Chen J. Continuous flow microfluidic device for cell separation, cell lysis and DNA purification. Anal Chim Acta 2007; 584(2): 237-43.

Index